MW01515011

ALSO BY CHRIS FERRIE

*Quantum Bullsh*t: How to Ruin Your Life
with Advice from Quantum Physics*

42 Reasons to Hate the Universe (And One Reason Not To)

*Where Did the Universe Come From? And Other Cosmic
Questions: Our Universe, from the Quantum to the Cosmos*

The Cat in the Box

Quantum Physics for Babies and many other books
in the bestselling Baby University series

Cosmic Bullsh*t

A Guide to the Galaxy's Worst Life Hacks

Chris Ferrie

sourcebooks

Copyright © 2025 by Chris Ferrie
Cover and internal design © 2025 by Sourcebooks
Cover design by Pete Garceau
Cover images © Aphelleon/Getty Images, GeorgePeters/Getty Images
Emoji art © streptococcus/Adobe Stock

Sourcebooks and the colophon are registered trademarks of Sourcebooks.

All rights reserved. No part of this book may be reproduced in any form or by
any electronic or mechanical means including information storage and retrieval
systems—except in the case of brief quotations embodied in critical articles or
reviews—without permission in writing from its publisher, Sourcebooks.

No part of this book may be used or reproduced in any manner for the
purpose of training artificial intelligence technologies or systems.

This publication is designed to provide accurate and authoritative information in regard
to the subject matter covered. It is sold with the understanding that the publisher is not
engaged in rendering legal, accounting, or other professional service. If legal advice
or other expert assistance is required, the services of a competent professional person
should be sought.—*From a Declaration of Principles Jointly Adopted by a Committee
of the American Bar Association and a Committee of Publishers and Associations*

References to internet websites (URLs) were accurate at the time of
writing. Neither the author nor Sourcebooks is responsible for URLs that
may have expired or changed since the manuscript was prepared.

Published by Sourcebooks
P.O. Box 4410, Naperville, Illinois 60567-4410
(630) 961-3900
sourcebooks.com

Cataloging-in-Publication Data is on file with the Library of Congress.

Printed and bound in the United States of America.
VP 10 9 8 7 6 5 4 3 2 1

To the universe

I hope you're not taking notes.

Contents

Preface

You are reading a book. It has an author. He is an Aquarius. If that seems like relevant information to you, you've picked up the right book because that fact is actually irrelevant bullshit. Worse, it's *cosmic* bullshit.

What is cosmic bullshit? In the most succinct way possible, I define it as *deceptively profound nontruth*. It's deceptive because the bullshitter is concealing that they want something from you. Falling for their bullshit could cost you money or time, but these days, the target is most likely your attention, the precious fuel of social media algorithms, which keep your eyes on the endless content served up between personalized ads. Someone pushing nontruth intentionally avoids honesty. They don't need to be a *liar*—after all, a liar understands enough to know what's true, while a bullshitter just cobbles words together. And when that's all they have to rely on, it's best to make it sound deep and important—in other words, *profound*.

What could be more profound than the depths of the entire cosmos? Wait. Hold that thought. I just got a notification that @universaltruthbeauty9123 has a new post.

Did you know we're all just STAR DUST? ☀️✨ *Yes, YOU, sitting there scrolling through your feed, are made of the same stuff as the stars! Mind-blowing, right?* 🪩 *So next time you feel down, just remember you're basically a walking galaxy.* 💫🌙 *Let's all shine bright together! Like if you agree you're celestial AF, share if you're feeling that cosmic connection, and comment with your star sign to see what the universe has in store for you!* 🌠🖤 *#CosmicVibes #StarDustSquad #UniverseWithinUs*

I just made that up, but I still want to hit "like" or "love" on it! We can't seem to help emotionally connecting to seemingly profound revelations that make us feel powerful in spite of our cosmic insignificance. But entertaining this bullshit can be as dangerous as mistaking a pharmaceutical ad for a public service announcement about free yoga in the park. From astrological signs to technological fantasies, ancient creation myths to UFO abductions, doomsday prophecies to time-travel paradoxes, misinformation and conspiracy theories abound in our world—even though we have more information and knowledge than ever before. In this book, we'll dismantle the most pervasive delusions sweeping our news feeds. In the pages to come, you can expect Albert Einstein's theory of relativity, quantum stuff, and the exacting tools of logic and reason. Just know that your comforting illusions may not survive the journey.

Debunking bullshit is an old tradition. Typically, pseudoscience skeptics focus on specific claims and purported evidence. I'm not going to waste your time doing that. Instead, I'm going to cut to the core of bullshit, cracking open its dry, crusty exterior to reveal why it really stinks. Humor will be attempted, insults will be thrown (in place of spears), and along the way, you will learn what truly makes the universe profoundly beautiful.

1

In the beginning...

Have you ever sat on your porch, sipping a hot coffee and staring at the distant sunrise while taking in all that's in between—the hills, trees, grass, birds, ants—and wondered how it all began, desperately hoping some old dude with a flowing white beard will pop up and impart the secrets of the universe on you? Have you ever done that? Well, it actually happened to me—except the coffee was lukewarm, and the dude was trying to sell me a vacuum cleaner. I asked him about the origin of the universe anyway, and, to my surprise, he had a theory. He leaned in, the scent of aged leather and moth balls blending with the morning dew. His eyes twinkled with the mischief of a man who knew where Jimmy Hoffa was buried. "All right," he began, his voice dropping to a conspiratorial whisper, as if the sparrows might be eavesdropping. "You really want to know how it all started?"

He spoke of a great Void—a perfect, spotless, cosmic emptiness maintained by the Cleaners, divine beings of perfection. The Cleaners grew weary of the monotony, so they spoke the sacred Word, and lo, it echoed through the emptiness: "Let there be Dust!" And Dust was, and it was good. For without Dust, what purpose have the Cleaners? And so the celestial dust settled upon the void, giving form to formlessness, tasks to the idle, mess to the pristine. The Cleaners saw the Dust and saw that it was...a job for later. But the Dust was more than it seemed. It contained the seeds of life itself, tiny specks of potential that clung to the fabric of space like pet hair on a carpet.

The Dust whirled and danced, colliding, coalescing, creating stars, planets, people, and homes with multiple flooring types. The Cleaners yearned for a return to eternal cleanliness and have bestowed it upon the chosen few who spread the news of the sacred instrument. He then picked up his vacuum, cradling it reverently before foretelling of the Return to Spotlessness. Shortly after, the man disappeared, leaving me on my porch with a now-cold cup of coffee and an extended warranty plan.

From ancient civilizations crafting tales of world-creating titans to modern-day vacuum salesmen with a pitch as expansive as the universe itself, humans have always tried to fill the void with stories. My vacuum salesman's story might have read as absurdly modern, but its essence is identical to the myths of yore—gods and heroes shaping the stars, carving the earth, and lighting the fires of existence. Each civilization, each era, spins a yarn to quell the existential dread of the unexplained origin of our existence. Within each culture, these stories become ingrained, a crucial part of each individual's conception of reality. While some will offer a modest sacrifice, such as a free set of steak knives, others have gone to war to defend their creation myths.

More than selling vacuum cleaners and bobblehead saviors for your dashboard, these tales give us the intangible thing we desire most: *control*. Control over our narrative, our beginning, and our personal destiny. Myths may offer solace, but through the lens of science, we glimpse the true magnificence of our universe. And while they may trade comfort for a bit of existential terror, the stories science tells are totally worth it.

It all started when...

Pick up any religious text, spiritual guide, or even a fictional narrative, and you'll start at the same place: *the beginning*.

We will probably never know what the first creation story was. Evidence such as cave paintings suggests that early humans had complex spiritual lives, and it is thus likely that they had creation stories. However, barring the archaeological find of the century, the specifics are just as likely lost to time. We do have some really old stories, though, so let's start there.

In southern Australia, the Gunditjmara people's creation story tells of an ancestral being called Budj Bim, which revealed itself as a volcano in the land. The lava and rock were its blood and teeth, which it spat across the earth to flow into the sea, creating new land to sustain the people. While many creation stories seem disconnected from actual events, this one appears to be based on truth. Budj Bim is indeed an extinct volcano that created the Tyrendarra lava flow that changed the land to create the uniquely habitable country in that region. But here's the thing. The last time Budj Bim erupted was thirty-seven thousand years ago. Pause for a moment and consider what this implies. This origin story has faithfully passed through over a thousand generations of people. It is likely the oldest surviving story still told.

This is, of course, not the only Indigenous Australian creation story, even among those stories that have survived. Before European colonization, hundreds of distinct communities passed down their own unique stories, rather than the single dominant cultural myth that exists on the continent today, which holds that all was created from the fermentation of malt and hops, which apparently requires ritual day drinking. In any case, common themes among the *original* Australian stories include local geological features as well as indigenous plants and animals.

The African continent, being the birthplace of humanity, has obviously been home to countless creation stories. These stories often have humanlike creator gods partaking in common earthly rituals that bring about the material world from a formless prior existence. Whereas, across the Atlantic, Native American creation stories often involve the emergence of our world from a previously existing one, such as when a sky god painted mud on the back of a giant turtle to create the Earth.

Oral traditions are often reserved for the cultures and communities that hold them, which, even when they survive, is why they are not widely known. There is, of course, the other obvious problem. In the children's game telephone, one person starts by whispering a word or phrase into the ear of the person next to them. The message is passed from person to person until everyone has heard it. The last person then says out loud what they heard, which, and certainly if the game involves boys, is usually *penis*. Luckily, adult humans don't resort to such juvenile humor. Written stories, on the other hand, have the unique ability to transcend the need for a living medium and are more resilient to pranksters.

Making cosmic omelets

The Enuma Elish, considered to be the oldest *written* creation story, is an ancient Babylonian myth that was etched on clay tablets long before the delete key and autocorrect. It is a cosmic-scale family drama of extreme pettiness, as if someone gave Kayne West his own reality TV show. Basically, one side of the family names a champion, Marduk, who turns out to be a real asshole by getting everyone drunk before having them agree to make him the overlord if he slays the primordial goddess, Tiamat, on the other side of the family, which he eventually does, using her body to create the world and also sacrificing her husband, Kingu, to create humans, who lived happily ever after. Phew.

Ancient Egyptians recorded several creation myths, usually found carved in the stone walls of the tombs of presumably important people. In one, the sun god Ra emerged spontaneously from the primordial waters to win gold in the solo synchronized swimming event, his tears of joy being used to create his adoring fans—that's us. In another tale, the god Atum was hatched from an egg, again emerging from the flooding primordial waters. You can start to sense a theme here.

The Rig Veda, one of the oldest Hindu scriptures in India, has numerous conflicting hymns dedicated to the origin of the universe. The most prominent recurring themes include the universe (or its creator) being hatched from an egg (usually golden) and the stuff of the universe (including humans) being made of the sacrifice of a primordial being. However, the Nasadiya Sukta hymn in the same text takes a skeptical stance with a bunch of poetic questions, such as, literally, "Who really knows?" Now, that's a church doctrine I can get behind.

The earliest writings of ancient China have similarly recorded many different creation narratives with now familiar motifs. Consider Pangu, a giant primordial being who emerged from... wait for it...an egg and split its body, which became different parts of the Earth. While water sometimes appears in these stories, the preferred basic cosmological element seems to have been the "life force" called *qi*. China brings novelty in the reverence of dualities like male and female, land and sky, hard and soft, Democrat and Republican, and so on, framing the concept of yin and yang, which probably has nothing to do with its depictions and use in the modern world. These dual elements are usually considered to have been one in primordial times and then split as the world emerged.

Since all of these stories are mutually contradictory, it should be clear that they can't all be true. Indeed, the true believer of one of them would have scoffed at the idea that any other was a factually accurate world history. With overwhelming likelihood, they are all false. Each story was believed not because of the lack of hearing of contradicting stories or evidence but because they made perfect sense within the context of the culture in which they were immersed. The diversity of stories doesn't point directly at human fallibility but to our need to explain our existence. As our understanding of and control over the world evolved, so too did culture and the stories that came along with it.

Going viral

In the Greek creation narrative, before the world as we know it took shape, there was only Chaos—with a capital C. In modern colloquial use, chaos is what ensues when an adult tries to build IKEA furniture without the instruction manual, or when a child is given IKEA furniture and a bunch of crayons. Capital-C Chaos is the initial ambiguous, formless, void state of the universe, which is more like the IKEA warehouse store you bought the furniture from...minus the Swedish meatballs.

From this nothingness popped Gaia—the personification of the Earth—and a few other deities representing real and abstract stuff. But Gaia, upon deciding that dating is overrated, miraculously conceived a son, Uranus, the sky god, whom she forced to become her booty call. Together, they created the Titans, monsters, and other mythical stuff. Foreshadowing the dangers of inbreeding, the youngest child, Cronus, castrated his father in an act that won him the title of "harvest god." If that wasn't enough, to avoid karma, he decided to eat his own children. Unfortunately, no one told him that karma's a bitch, and Zeus, the slippery little runt, dodged the feeding and overthrew him. To make a long story short, cue the Olympians, with Zeus as the new boss—same as the old boss—and the continuation of endless shenanigans.

The psychological reflection in Greek mythology captured the collective unconscious of not only the Greeks but also all who followed by providing a mirror to the human psyche. Bluntly, it offered an endless source of entertainment. The stories, which continually played out rather than ending at creation, were the blockbusters of their day, gripping the populace with their epic scale and personal dramas. This shared entertainment experience fostered social cohesion, unifying the populace with a common lexicon of symbols and narratives. While that seems quaint, let's not forget the true villains of any story—politicians. Creation myths were employed by leaders to reinforce social order and justify their rule, often claiming divine favor or descent.

In a stroke of not-so-cunning cultural appropriation, the Romans, with their renowned "originality" in the arts, copied the Greek pantheon wholesale, repackaging Greek gods with new, sexier Latin names like Apollo and Venus. It was through Roman conquests and each empire's love for all things Greek that these myths became the shared heritage of many civilizations we still enjoy in art and entertainment today.

I can haz lite?

The Greek mythical soap operas, with their divine debauchery and celestial catfights, laid the groundwork for the more solemn, monotheistic narratives of the Abrahamic religions. These newer narratives traded the anything-goes Greek dramaturgy for a more...let's say, streamlined plot with a definitive endgame, thus ensuring their own brand of psychological and political sway that persists to the modern day.

The creation story from the book of Genesis, which eventually became part of several distinct religious texts, details the most well-known creation story. Variously depicted as an old white dude with a glorious beard and unnecessarily sick abs, the main character, God, creates the world in six days and rests on the seventh. While it doesn't really explain why it took him so long, God simply speaks into existence light, sky, land, celestial bodies, animals, and, finally, humans. Since it definitely wasn't originally written in English, one translation is just as good as another. Let's go with the official "lolcat" version, and I quoteth:

Genesis 1:1–3. Oh hai. In teh beginnin' Ceiling Cat maded teh skiez An da Urfs, but he did not eated dem. Da Urfs no had shapez An haded dark face, An Ceiling Cat rode invisible bike over teh waterz. At start, no has lyte. An Ceiling Cat sayz, i can haz lite? An lite wuz.

You know when you are snooping around in your hotel room with a black light, and you open the bedside drawer to find a book with glowing smears and splatters on it? That's the King James Version of the Bible. Published in 1611, it remains to this day the most printed book of all time, which is why many know the story even if they haven't read it themselves. That being said, some read it a little bit *too* literally. Stay tuned for when these assholes make their entrance in our mostly chronological tale.

Creation story Mad Libs

Let's pause for a moment and take stock. Mulling over all the creation myths I've outlined, detailed, and mocked, can you find anything common to all of them?

Probably the most important aspect of any creation story is the fact that it has to deal with "the before." Most start with some primordial state, often described as a void from which everything else emerges. True to our desperate attempt to lay blame for everything that happens, creation is often attributed to some anthropomorphized being, or beings, who shape the universe and its contents. The process itself involves first bringing order to chaos, establishing a rigid and tangible stage upon which the rest of the story is set—think sky, earth, water, rock, and so on.

Next comes the main event: life. The creation of life, particularly humans, is always the climactic theme. (You'd almost think we just made this shit up out of hubris.) Life itself is often created from the body of a primordial being. After that, the primordial beings usually stick around in this ultimate bait-and-switch to force moral, existential, or philosophical lessons on us.

Now that you know the generic formula for a good creation story, let's make our own—it's Creation Story Mad Libs!

Once upon a time, in the vast expanse of _____ (noun), there existed a Primordial _____ (noun). Driven by a cosmic _____ (noun), it began to weave the fabric of _____ (noun). From its essence emerged divine _____ (plural noun), each embodying a unique aspect of _____ (noun). They danced to the tune of _____ (plural noun), shaping _____ (noun), _____ (noun), and _____ (noun).

Among the realms they forged was _____ (noun), a realm destined to harbor _____ (plural noun). The first beings were molded from the _____ (plural noun), gifted with the spark of _____ (noun). They were entrusted with the legacy of the _____ (noun), a reflection of the divine in a _____ (adjective) coil. As time _____ (verb (past tense)), civilizations _____ (verb (past tense)) and _____ (verb (past tense)), each striving to unravel the mysteries of their _____ (noun), reaching for the _____ (plural noun) from whence they came. Yet, amid the quest for _____ (noun), the whimsy of _____ (noun) entwined their destinies, an eternal dance of _____ (noun) and _____ (noun).

Sounds fun! I'll go first.

> Once upon a time, in the vast expanse of spaghetti **(noun)**, there existed a Primordial Chicken **(noun)**. Driven by a cosmic accordion **(noun)**, it began to weave the fabric of socks **(noun)**. From its essence emerged divine toasters **(plural noun)**, each embodying a unique aspect of cheese **(noun)**. They danced to the tune of tacos **(plural noun)**, shaping glitter **(noun)**, lunchtime **(noun)**, and eyebrows **(noun)**.
>
> Among the realms they forged was Bathtub **(noun)**, a realm destined to harbor rubber ducks **(plural noun)**. The first beings were molded from the crayons **(plural noun)**, gifted with the spark of jazz hands **(noun)**. They were entrusted with the legacy of the bubble wrap **(noun)**, a reflection of the divine in a squishy **(adjective)** coil. As time wobbled **(verb (past tense))**, civilizations burped **(verb (past tense))** and tangoed **(verb (past tense))**, each striving to unravel the mysteries of their leftovers **(noun)**, reaching for the Moon Pies **(plural noun)** from whence they came. Yet, amid the quest for nachos **(noun)**, the whimsy of kazoo **(noun)** entwined their destinies, an eternal dance of pickles **(noun)** and disco **(noun)**.

Righteousness redefined

From the blood-soaked Crusades to the explosive Thirty Years' War, faith has not only been a banner under which wars were waged but also a justification for them. While these conflicts weren't specifically about creation myths, they demonstrate the depth of influence religious beliefs can have, as major religions are often synonymous with their creator gods and origin narratives. In this context, the idea of *creationism* emerges as a significant aspect of religious belief, particularly in the Abrahamic traditions. Creationism, broadly speaking, is the religious belief that a god, as described by religious texts or those who claim to interpret them, created the universe and, more importantly, life—specifically, humans and, preferably, humans that look like the religious leaders. This belief system—from literal interpretations to more metaphorical understandings—has been central to many religious teachings and traditions.

The Creationist heydays were the so-called Dark Ages, aptly named because it was literally dark all the time. Kidding! It was only dark half the time...and metaphorically dark the rest. In Europe, the Church held the monopoly on the big ol' book of light. Basically, the Church was like that one friend who refused to download map data before driving through the Cellular Dead Zone (a.k.a. Montana), insisted they knew the way, and refused to stop and ask for directions when it became clear they were actually lost. The Church had a stranglehold on knowledge, education, and, most importantly, the narrative. Creation wasn't up for debate—it was a divine line everyone was expected to toe. Question authority, you say? This was 1023, not 2023. Speaking up wasn't a Sunday afternoon protest with witty homemade signs; it was a one-way ticket to the business end of a pyre.

The Dark Ages were a time when the Earth was flat if the Church said so, the sun revolved around us if the Church said so, and that woman with the "God Hates Flags" sign was definitely going to float in that lake if the Church said so. The motivation behind the narrative was less about understanding the world and more about controlling its inhabitants. Any creation story that didn't line up with the Genesis account was rooted out and not so politely shown the door. However, it wasn't all doom, gloom, and heresy trials.

And now, here's a message from our sponsor.

Tired of all this pestilence? Sick of working the field and paying taxes? Want some food and stuff? Come, abandon society, and devote your life to sobriety, celibacy, and making books no one will read for several centuries. Join your local monastic order today!

As the Dark Ages labored on, monks became the unlikely torchbearers of knowledge, meticulously copying ancient texts and manuscripts, setting the stage for an unknown future where people could think for themselves without worrying about the latest divine innovation in torture. Talk about playing the long game! The next time you enjoy practically anything in the modern world besides sex and drugs, be sure to thank the nearest monk.

Seeing the light

As the Renaissance dawned, the Church's grip loosened, and the only things spreading faster than the plague were ideas. In the musty halls of academia, where scholars previously busied themselves determining how many angels could dance on the head of a pin, whispers of knowledge could be heard on the winds of change. This was the era of the Renaissance man, not to be confused with the modern Renaissance man, who is just a guy "reaching out" on LinkedIn, perpetually ready to "jump on a call" with you. Enter Galileo Galilei, the seventeenth-century version of a thought leader, armed with nothing but a telescope, a sick lute, and a dangerous idea.

Though Galileo didn't invent the telescope (or the lute), he significantly improved the telescope and had the genius idea to point it toward the heavens. Through his mighty instrument, he discovered moons orbiting Jupiter, phases of Venus, and the rough surface of the Moon—observations that were about as welcome in the Church's court as a Kia hybrid at a monster truck rally. These findings challenged the prevailing views of perfect heavens revolving around an Earth-centered universe. What was more, Galilei invited anyone unconvinced to see for themselves—a generosity not reciprocated by the Church, though they were more civilized in their punishment this time.

The party moved to Europe's salons and coffee houses as the Renaissance launched into the Enlightenment. Imagine them as the IRL equivalent of Twitter, but with less dumpster fire and more powdered wigs. Here, amid debates and caffeine, the seeds of modern science were sown. In 1687, a book hit the shelves that would forever change how we view the universe. Isaac Newton's *Principia* included his famous three laws of motion and the universal law of gravitation, not to mention the invention of calculus. By the nineteenth century, Newton's theories weren't just part of science; they were the backbone of it. His influence was so pervasive that it set the scientific agenda for centuries.

Newton also served as an exemplar of reason and rationality

whose application defined the Enlightenment and its intellectual spoils. In addition to several other advancements in science and technology, we began to understand diseases as more than just angry spirits, that economies bowed to logical theories, and ideas of liberty and democracy might not be as bad as the king said they were. While many thinkers, including Newton, still maintained a belief in a creator god, there was a growing trend to seek explanations for the workings of the world that did not rely solely on the supernatural or miraculous. This shift represented a fundamental change in how we understood the world. It was no longer enough to simply describe. Now we had to quantify, predict, control, and replicate. But to really make headway, we needed to get rid of this pesky "God" character...

All-natural selection

In the 1830s, a young naturalist journeyed to the far corners of the Earth, where he observed the diverse tapestry of life, reflected on the origins of the wide array of species, and finally concluded he was related to monkeys. That young naturalist was named Charles Darwin. The primary purpose of his famed HMS *Beagle* voyage, under the command of Captain Robert FitzRoy, was not to rewrite history or question the pedigree of aristocrats but to survey the coast of South America, which was crucial for British maritime interests at a time when British maritime interests included owning literally everything.

Darwin was invited to join the expedition as a naturalist, but the role was not part of the ship's official complement. In fact, he was to be a gentleman companion for Captain FitzRoy, who sought a person of similar social standing and education to accompany him, lest he have to talk to a boat full of knaves, scabs, and scallywags. Darwin seized the opportunity and took as his role to collect, study, and document various natural specimens, including plants, animals, rocks, and fossils. In case you were wondering how a "naturalist" collects animal specimens, it usually involves a gun and a jar of preserving alcohol.

Over the five-year voyage, Darwin meticulously observed and cataloged the natural world, from the diverse species of the Galápagos Islands to the geological formations of South America. But it wasn't just a grand adventure involving two most-elite gentlemen and their motley crew. No, it was a scientific odyssey that culminated in one of the most revolutionary ideas in human history—*the theory of evolution by natural selection.*

In 1859, Darwin published *On the Origin of Species,* a book that would rival Newton's *Principia* for the most influential text of all time. Darwin proposed that species evolve over time through a process of natural selection, whereby individual organisms with heritable traits best suited to their environment are more likely to survive, reproduce, and hence pass on those traits. This process, over a vast period of time, leads to the gradual emergence of new species. Darwin's idea was a radical departure from the prevailing view of unchanging species, each created separately and designed for a divine purpose...except wasps—fuck those things.

Darwin's work was revolutionary not only because it provided a simple and compelling explanation for the diversity of life on Earth but also because it suggested a common ancestry for all living organisms. This obviously offended a lot of people. Samuel Wilberforce, the Bishop of Oxford, famously asked Darwin's colleague, "Was it through his grandfather or his grandmother that he traced his descent from an ape?" The reply was long-winded and pretentiously typical of British academic debates, but a modern translation would be, "Ask your mom."

Textbook science

Most of us learned about evolution the same way—from that scene in *Jurassic Park* where scientists make dinosaurs in the lab just before the lawyer gets eaten by a *T. rex*. It's got something to do with DNA, right? For those with parents who aren't clutching their pearls as tightly as their Bibles, you probably got your first lesson on evolution in elementary school science class. You were probably shown that same silly caricature of an ape transitioning to a man as he walks left to right. But this is not very accurate.

To be fair, Darwin himself didn't have an accurate idea of the mechanisms behind evolution either—the concept of *genes*, let alone DNA, didn't exist back then, for example—which is why children don't hear about Darwin himself much. The modern understanding of evolution is often called the *modern synthesis*, which combines Darwin's theory of natural selection with modern genetics. Basically, it works as follows.

Every organism carries genetic information in the form of long molecules called DNA—deoxyribonucleic acid for those in lab coats. DNA is microscopic and doesn't directly cause the differences we see in individual organisms. However, it does encode visually distinct heritable traits by dictating the structure and function of *proteins*. Proteins perform direct functions, such as being the structural building blocks of cells and tissues, but they also do so much more. They act as tiny machines—switches, pumps, motors, and vehicles—catalyzing chemical reactions and transporting molecules around the organism. Each section of the DNA molecule, called a *gene*, is a blueprint for a specific protein, and variations in these genes can lead to changes in the proteins produced, thus influencing physical traits.

In the wild, resources like food, water, shelter, and influencer status are limited. This scarcity leads to a struggle for survival. Organisms with traits that give them an advantage in their environment are more likely to survive longer. The longer an organism survives, the more time it has for...ahem...well, you know. You carry two near-identical copies of your DNA, one inherited from each parent, and much of this genetic material is shared across all humans to encode the essential aspects of human biology.

This shared genetic information is responsible for the basic human ingredients such as two arms, two legs, ten fingers, ten toes, a pair of eyes, a nose, a mouth, two ears, a set of thirty-two teeth, a heart that beats rhythmically, a pair of lungs for breathing, a stomach and intestines for digestion, a liver and kidneys for detoxification and waste removal, a complex network of blood vessels, a skeletal system providing structure and protection, muscles that enable movement, skin that acts as a protective barrier, a reproductive system, a network of nerves for sensing and responding to the environment, glands that secrete essential hormones, an immune system to fight off pathogens, hair covering various parts of the body, and, for everyone but "undecided voters," a brain.

The nonidentical part encodes the differences between your parents, which combine to create your individual traits. During reproduction, you make a near-perfect copy of your DNA, adding a light sprinkling of variation in the genes, eventually contributing half to your offspring. Their new mix of genes might include variations in eye color, hair type, and other inconsequential things for non-gingers while hopefully maintaining standard human features determined by genes all humans share. Some traits might confer an advantage over other humans competing for a mate, and (cue *The Lion King* intro music) the cycle continues—though we typically like to view our species as above the animalistic behaviors that necessitate a David Attenborough narration.

While Charles Darwin didn't understand the molecular details, his observation of trait inheritance and natural selection laid the groundwork for modern evolutionary biology. Understanding the mechanism only bolsters the conclusion. Over time, as populations of organisms accumulate different genetic variations, they diverge from each other, leading to the formation of things so different we call them new species. This process is not linear, but like branches splitting off from a main trunk, each heading in its own unique direction—creating the so-called *tree of life*.

This isn't just a metaphor. It's a scientific representation of the evolutionary relationships among all life forms on Earth, tracing back to a common ancestor. Darwin's groundbreaking idea of common ancestry suggests that if you run the clock backward, you'd find that all living organisms—from humans to hydrangeas to hummingbirds—share a common root. This tree has branched out over billions of years, with each branch representing a different lineage. In short, Darwin did not descend from an ape, but we and apes have a common ancestor, probably less impressive than either of us. In other words, we and the apes are cousins—if that pill is easier to swallow.

Unnatural selection

That little story is how evolution works for most of the living things we can see with our eyes. Ancient creation myths obviously don't refer to things as weird as viruses, some of which don't have DNA and reproduce via cloning themselves instead of sex. But, irrespective of the details of asexual reproduction, which I am hesitant to make a porn-related joke about since I haven't checked the internet for it yet, Darwin's generic idea still applies. So long as there is a population of things that reproduce with variations that are selected based on environmental fitness, *evolution* will happen. In fact, computer scientists use this template to solve extremely difficult problems by *mimicking* evolution by natural selection—they even call it *genetic programming*! Much as in nature, programs "compete" by having their solutions compared. The best programs "survive" and clone new programs with slight variations, and (cue *The Lion King* intro music) the cycle continues.

The concept of *artificial* selection—where humans, rather than nature, select for desirable traits—serves as a powerful testament to and extension of Darwin's idea. This human-guided evolution, seen in everything from agricultural practices to algorithmic simulations, doesn't invalidate Darwin's ideas. Instead, it bolsters them by showcasing the principles of evolution in a controlled setting. Indeed, Darwin himself used the breeding of plants and animals as an analogy for his theory.

For thousands of years, humans have been practicing artificial selection in agriculture. By choosing plants and animals with preferred traits for breeding, we have dramatically altered species to suit our needs. For example, corn is the product of selective breeding of teosinte, which looks more like what comes *after* corn when your dog has eaten too much of it. Speaking of which, from Chihuahuas to Great Danes, dogs are another testament to the power (or failure, from the Chihuahua's point of view) of artificial selection. While nature might take millennia to shape a wolf, human-guided breeding can turn a wolf into a fashion accessory within a few generations. This showcases the high-level mechanisms of heredity and variation that Darwin first proposed, not to mention the utter cruelty implied by "survival of the fittest"—be it in the wild or in a handbag.

When someone wants a black-and-white dog, they take a black dog and a white dog and...you know, hope for the best. The point is, selective breeding, which does not require knowledge of modern microbiology, is still largely a "random" process. Whereas today we can create living things with exactly the traits we want through *genetic engineering*. Doing this with mammals is fraught with ethical hurdles, but if history is a good guide to what humans will do in the future, which it usually is, we will soon be able to decide the exact number of spots Spot will have. I'd choose just one.

While it's not commonplace for animals, we genetically engineer plants all the time. The newest types of corn, for example, are not bred so much as they are designed, creating a Whole Foods shopper's worst nightmare: a GMO. Anything that has had its DNA artificially altered is referred to as a *genetically modified organism*. If conventional selective breeding speeds up the race to a desired trait, genetic engineering uses the cheat code.

With the ever-increasing power of science, the idea of passing off the responsible use of it to future generations becomes more difficult, regardless of the ease with which boomers seem to have done it in the past. As we wield the ability to shape life itself, ethical considerations come to the fore. Where do we draw the line in genetic manipulation, what havoc might we wreak, and what responsibilities do we have toward the species we alter? Recall the foreshadowing of Jeff Goldblum as Ian Malcolm in *Jurassic Park*: "Your scientists were so preoccupied with whether or not they could that they didn't stop to think if they should." To be fair, though, the most dangerous consequence of genetic engineering in that universe seems only to be too many sequels starring Chris Pratt.

When phrased negatively, the answer to "Should we make new dinosaurs?" seems obvious. But we could also ask, "What happens if we don't do it?" Do we have a responsibility to treat life-altering genetic conditions or correct genetic disorders before a child is even born? On a larger scale, genetic technologies could play a role in biodiversity conservation, perhaps resurrecting extinct species or bolstering the genetic diversity of endangered ones. If we can shape the world for the better, should we?

The fact that we can even consider the ethical implications of our newfound powers is, in itself, a testament to Darwin. Each ethical dilemma, each decision about what to preserve, change, or enhance, reaffirms the depth of our scientific understanding of the essence of life. Our ability to manipulate life at a genetic level is a powerful reminder of the evolutionary mechanisms we now control.

Our awesome capabilities in genetics and artificial selection don't just extend Darwin's ideas but also serve to underscore the validity of evolutionary theory, challenging traditional creation myths. These myths, while culturally significant and rich in symbolism, remain static and confined to their times. In contrast, Darwin's theory, continually supported and expanded upon by modern science, is embedded in a framework that adapts and grows with each new discovery. Rather than a conviction that, at best, provides intellectually numbing comfort, science is a commitment to the continual expansion of our understanding. This fundamental difference in conviction is what propels us forward, fueling the relentless pursuit of answers.

But questions like those about the origin of life cannot be answered with a single science alone, so our story must move on. Yet, as we transition beyond Darwin *forward* in time, we ironically also move backward.

Where did the Earth come from?

We can't say definitively how and when life began. However, we've got a pretty solid grip on the age of our cosmic neighborhood. The Earth, Sun, and the rest of the gang in our solar system are roughly 4.5 billion years old. And this isn't just an educated guess—it's a figure backed by a chorus of scientific disciplines, primarily geology and astrophysics. And it directly contradicts many of our favorite creation myths.

The process of dating our planet and its solar siblings involves a bit of nuclear detective work. Scientists can analyze the decay of radioactive materials in rocks, minerals, and even leftover pizza to calculate their age. This decay happens at a rate as predictable as the demands of a Swiss train passenger, with each radioactive material having an internal clock known as its *half-life*. Imagine an hourglass, but instead of sand, it's filled with atoms falling apart, with half of them decayed in one half-life. So, by measuring the amount of decayed material in a sample and knowing its half-life, we can figure out its age.

Carbon dating has been the superstar of the dating world since the 1940s. It's been the go-to method for dating organic materials that are less than fifty thousand years old, reshaping fields like archaeology, paleontology, and climate science—a real science, despite what that guy screaming on YouTube said. However, the limitations in dating older materials necessitated the development of other methods for a deeper dive into Earth's history.

To estimate the age of the Earth, scientists rely on uranium-lead and rubidium-strontium dating methods, which don't quite roll off the tongue as well as carbon dating. However, while the type of carbon used in carbon dating has a half-life of a few thousand years, these other minerals have lifespans in the *billions*, allowing them to track longer historical timescales. When applied to ancient rocks, meteorites, and even moon rocks, these methods have consistently indicated that the Earth and the rest of the solar system are about 4.54 billion years old.

Meanwhile, astrophysicists have provided estimates that corroborate these geological findings, reinforcing a cohesive picture of the solar system's formation and evolution. Astrophysical estimates of the age of the solar system come from the study of stellar evolution. The Sun is what we call a *main sequence* star. The main sequence phase of a star is defined by the stable burning of hydrogen into helium in its core. A star's lifespan in the main sequence largely depends on its mass. Larger stars burn through their hydrogen fuel more quickly and have shorter main sequence lifespans, whereas smaller stars, which burn their fuel more slowly, can spend much longer in the main sequence. Given the Sun is 1 solar mass (not a coincidence), roughly a medium-sized star, our best solar models predict a total main sequence lifespan of about 10 billion years.

Scientists use astronomical observations, theoretical models, and computer simulations to understand the life cycle of stars. By measuring a star's brightness (technically, *luminosity*), surface temperature, and composition (which is inferred from its light spectrum), we can estimate its current age. For our Sun, these characteristics are consistent with a star that is right in the middle of its main sequence phase. The guess is not as precise as those from the rock collectors, but it is in the ballpark of 4.6 billion years, which is consistent with independent estimates. This convergence of evidence from multiple disciplines highlights the robustness of the current scientific understanding of the age of our solar system.

We also have a pretty good understanding of *how* the Earth came to be. Astronomers have observed regions in space known as *stellar nurseries*—essentially clouds of mostly hydrogen and other elements left over from the supernova explosion of former stars. Regardless of what they'll tell you in Nashville, these cosmic dust clumps are where real stars are born. In these nurseries, *protostars*—those in the early stage of star formation—are often surrounded by rotating disks of gas and dust, known as *accretion* disks. These disks are the sites where planets, asteroids, and other solar system floaties form. Basically, the Earth was made from the leftover materials that didn't make the cut for stardom. As the Sun ignited in its stellar nursery, the residual dust and gas swirling around it gradually coalesced under gravity's unyielding grip.

Over time, these particles stuck together, forming larger and larger clumps, much like how a snowball grows as you roll it down a hill, except with less frostbite and more cosmic significance. When the snowball gets big enough and clears enough stuff, it becomes a planet—that's us.

Interestingly, almost every atom that exists on Earth today was around back then. Mountains, water, air, soil, you, and that pile of shit are all just those same atoms arranged in different complex patterns. Those atoms were the product of a supernova explosion that formed the stellar nursery in which our Sun and Earth formed. So, in a way, you're made of stardust—a fact that's both truly humbling and a bit of a cliché for science documentaries. You're a curious mix of the laws of nature grinding away while random flukes in the arrangements of atoms eventually coalesced into a pattern complex enough to replicate itself.

For an ancient human, this origin story would suffice, as they would not have known that the stars are other Suns, that the Milky Way was a galaxy of hundreds of billions of them, and that there are trillions of other galaxies waiting to be found. Now, with the aid of science and technology, we know there must be more to this story, for where did our galaxy come from...and what came before that?

Space, time, and space-time

With few exceptions, most cultural myths have a creation component. Though specific dates aren't usually mentioned, the universe, for many cultures, certainly had a beginning. The opposite has been true for most of the history of physics. The Greek philosopher Aristotle is probably the most influential physicist in history. Aristotle envisioned an eternal cosmos as a series of concentric, crystalline spheres revolving around a stationary Earth, with stars fixed on the outermost sphere. This geocentric model, while incorrect, was the backdrop of cosmological thought from the time of writing around 300 BCE until the Scientific Revolution. Though much of his physics was speculative or just plain wrong, his idea of an eternal universe survived. We might even say it was strengthened.

As discussed a few pages back, Newton revolutionized our understanding of the physical universe through his mathematization of space and time. He introduced the concepts of absolute space and absolute time. He envisioned space as a vast container within which objects exist and move and time as a flow that marches forward, unaffected by the events within it. While not a Russian nesting doll of crystal spheres, this clockwork universe was an infinite and static stage upon which the laws of motion turned the gears within.

Yet, there was always a sense of unease. Newton, though he was certainly comfortable with nearly blinding himself by sticking a needle behind his eye, admitted the idea of gravity acting at a distance through empty space was philosophically unsatisfying. His famous words, "I frame no hypothesis" for the mechanism of gravity, foreshadowed a revolution in our understanding of space and time that would tear down the entire edifice of classical cosmology.

In 1905, the most pivotal year in the annals of physics, a young German patent clerk working in Switzerland published several papers on independent topics that would forever change our understanding of the universe, and what a Google image search of "scientist" should return. Albert Einstein, then unknown in the world of physics—and with only a little exaggeration—proved the existence of atoms by inventing statistical physics, proved the existence of photons (atoms of light) by inventing quantum physics, and disproved the existence of absolute space and time by inventing relativity. While statistical physics is complex and quantum physics is confounding, relativity is a real mind-bender.

Einstein considered *inertial* observers—those moving at a constant speed and direction. Think standing still or sitting on a train. He then made two assumptions. The first assumption was that the laws of physics are the same for every observer, which is not controversial. The second, which is less intuitive, is that the speed of light will be measured to have the same value for every observer. This speed, which is 300 million meters per second, is usually denoted with the letter c. Together, these two simple statements comprise the *special* theory of relativity. It's "special" because of the inertial observer thing, not because it was Einstein's favorite little theory that could do no wrong, and all of its consequences went straight onto the refrigerator door.

While it's not obvious, special relativity directly implies the radical notion that space and time are not separate entities but must be codependent, combined into something we now call *space-time*. There is no absolute space and no absolute time. According to Einstein's theory, events that appear to happen at the same time from one perspective may not be simultaneous from another if the observers are moving relative to each other. Einstein spent several years trying to remove the "special" in special relativity. He finally succeeded in 1915, allowing in his theory for observers to move in not just steady straight lines. The "nonspecial" version is called the *general* theory of relativity. The key novelty is the necessity of *curved* space-time, which interacts intimately with mass and energy.

Massive objects like stars and planets warp the space around them, and this curvature guides the motion of other objects. A classic picture to have in mind is variously sized balls on a trampoline. The balls move along the surface of the stretched fabric not because a force is pulling them together but because the surface itself guides them. Space-time is similar but has more dimensions, and the curvature-guided motion dictated by Einstein's theory is exactly what would be expected *had* a force (a.k.a. gravity) been there. In other words, Newton's gravity is an illusion: there is only matter and space-time. As John Archibald Wheeler put it, "Space-time tells matter how to move; matter tells space-time how to curve."

Einstein's insights transformed our understanding of the universe from a static, unchanging stage to a dynamic, evolving entity. His theories suggested that the universe is not a fixed backdrop but a dynamic arena where space-time ebbs and flows. This shift from the Newtonian view of a static, unchanging universe to a dynamic, evolving cosmos is one of the most profound transformations in our understanding of the universe, and it set the stage for modern cosmology, framing our current scientific understanding of the universe's origins.

An expanding universe

Initially, Einstein favored a static space-time model, at least on cosmological scales. He was so entrenched in this view that he introduced the *cosmological constant*, an extra term in his equations, to ensure a static universe. Einstein knew this was a theoretical crutch, but the prevailing scientific consensus of the day couldn't conceive of an evolving cosmos. Little did he know, his own equations held the seeds of a revolutionary idea.

Alexander Friedmann, a Russian physicist and mathematician, and Georges Lemaître, a Belgian physics professor, were working independently on Einstein's equations in the 1920s when they discovered solutions that contained an expanding universe. Friedmann's work, starting in 1922, revealed a range of possibilities where the universe could expand, contract, or do a bit of both, much like the waistlines of every middle-aged dad during the holiday season. Meanwhile, Lemaître, in 1927, proposed a similar model but went a step further, suggesting the universe began with the explosion of a "primeval atom." The allusion to religious mythology was probably not coincidental, as Lemaître also happened to be a Catholic priest. Einstein, upon first encountering Lemaître's work, famously dismissed it, remarking, "Your calculations are correct, but your physics is abominable." This skepticism, however, soon gave way to acceptance as observational evidence began to mount.

Even late into the 1920s, the prevailing view was that our Milky Way galaxy constituted the entire universe. This cozy, somewhat provincial view of the universe was akin to the simple geocentric model of centuries prior. Unlike Galileo's backyard telescope, Edwin Hubble had at his disposal the world's largest telescope, the 2.5-meter Hooker telescope stationed at the top of a California mountain. But, similar to Galileo, Hubble observed various "nebulae"—thought to be clouds of dust in the Milky Way—and discovered they were entire galaxies outside the Milky

Way. We went from being the center of a very tiny universe to being in a completely unremarkable spot in a typical galaxy in a cosmos vast beyond comprehension.

Surprisingly, this wasn't even Hubble's most famous discovery. While studying these distant galaxies, he discovered that they were moving away from us, and the farther away they were, the faster they seemed to be receding. Hubble concluded that the universe is expanding, and it does so uniformly in every direction. This was the observational evidence that corroborated Friedmann's and Lemaître's theoretical models of an expanding universe. Einstein eventually conceded that his introduction of the cosmological constant was unnecessary, calling it his "biggest blunder."

Now, if everything is moving away from everything else *forward* in time, what happens if we imagine moving backward in time? Well, obviously, everything must have been squashed up together at a single point, just as Lemaître suggested. In a 1949 BBC radio interview, physicist Fred Hoyle dismissively referred to the idea of a universe beginning with a gigantic explosion as the "big bang." His intention was to contrast Lemaître's dramatic and singular event with his "steady state" theory of a universe of constant matter creation maintaining a static density. Unfortunately for Hoyle, by giving it a catchy and memorable name, Lemaître's theory became ever-popular and is now essentially the accepted scientific model for the cosmos. Let that be a lesson for all the Hoyles out there. Don't be clever; just call them an asshole and be done with it.

The briefest history of time: a science origin story

In the beginning, our universe began not with a divine word or a cosmic dream but with a singular event: the big bang. About 13.8 billion years ago, the universe as we know it started with an incomprehensibly dense and hot tiny ball of pure energy. The big bang was not an explosion in space but rather a rapid expansion of space itself.

In the tiniest fraction of the first second, the universe underwent a period of rapid inflation. This exponential growth of space smoothed out and flattened the cosmos, much like taking a crumpled-up tablecloth and stretching it out quickly so that it appears flat and smooth. As the universe cooled and expanded, it reached temperatures conducive to nuclear reactions, and the first simple elements, primarily hydrogen and helium, began to form a still-hot plasma soup. Expansion and cooling continued until proper neutral atoms (protons with captured electrons) formed. As these atoms minimized their energy, they released light into the now-transparent universe.

This light has been traveling throughout the universe for the last 13.8 billion years, and we can still detect it today as low-energy radiation arriving at us from every direction. This *cosmic microwave background* is the oldest direct observational data we have of the universe, a map of the cosmos at its first few hundred thousand years of existence. A few hundred *million* years later, the universe had cooled enough for the first stars to form, igniting within clouds of hydrogen gas that pulled themselves together through their mutual warping of space-time. These stars were the nuclear furnaces where heavier elements were forged, elements that would one day become the building blocks of planets and life.

As galaxies formed and evolved, there was a mysterious invisible mass holding them together, which we call *dark matter*.

Though directly undetectable, we infer its presence from otherwise unexplained gravitational effects. Over the course of billions of years, dark matter has acted as the cosmic scaffolding on which galaxies and galaxy clusters were built. Coalescing into the current large-scale structure of the observable universe, we find a cosmic web of galaxy clusters intricately connected by filaments of other galaxies attracted to tendrils of dark matter.

Gravity ought to have been pulling all this mass, dark or otherwise, together. Yet, our universe is not only expanding but doing so at an accelerating rate. We don't exactly know what's causing this, as we cannot directly observe or measure it, but some mostly uniform source of outward pressure must permeate the universe. We call it *dark energy*, an obvious analogy to dark matter. Where it will take us, we don't yet know. But, here we are, a small if not wonderfully complex fluke of replicating information, in an ever-expanding universe filled with a vast array of galaxies, stars, planets, dust, and energy. This is our story, and while it doesn't place us at the center, it's still the best one we've got.

A new religion

While most discussions that feature both science and religion pit them against each other, it's interesting to note the similarities. First, both the scientific and mythological creation stories share a narrative structure that begins with a primal state—the singularity in the big bang theory or the primordial chaos in many myths. The story then progresses through a series of transformative events leading to the current state of the world, the culmination being, of course, the existence of a group of people asking the very question of how they got there.

Indeed, both types of stories seek to explain the origins of life, Earth, and the universe, aiming to address fundamental human questions about where we come from and how the world around us came to be. These existential questions seem to set us apart from all other living things in an ironic twist of evolutionary fate. The uncertainty can cripple us, paralyzing our minds which demand answers. So, it is no surprise that those same minds answer the call.

While past minds answered it with wars, in more recent times, the debate over creationism has shifted from battlefields to classrooms, courtrooms, and a shithole off Interstate 275. The Creation Museum in Kentucky, founded by a key player in the *young Earth creationism* movement, features dragon exhibits alongside depictions of humans and dinosaurs coexisting. It's like a *Flintstones* episode but played straight-faced. Such nonsense is necessary if one accepts a literal interpretation of the Bible. Young Earth creationists, basing their ideas on the lineage of Bible characters, reckon the entire universe is less than six thousand years old and was created in six days.

Then there's *intelligent design*, which is like creationism's bad attempt at a science fair project. It argues for an "intelligent cause" behind life's complexity, distancing itself from explicit biblical references but often sharing the same underlying motives of the more extreme movements. Cut to Dover, Pennsylvania, in 2004, where the local school board tried to slide intelligent design into the biology curriculum as if they were sneaking alternative facts into a Wikipedia article. Thankfully, only a year later, Judge John E. Jones III (appointed by President George W. Bush, no less) said, "Let's see who you really are," and pulled the lab coat off intelligent design in a *Mystery Inc.*-style climax to reveal it was creationism all along. Basically, teaching it in public schools is as unconstitutional as taking away the teacher's guns.

While the temptation is there to join the chorus of scientists picking apart religious beliefs with the precision of a neurosurgeon, that's not the rabbit hole we're tumbling down here. This isn't about taking a sledgehammer to Aunt Edna's cherished belief in a divine garden party starring Adam, Eve, and a conniving serpent. No, my task is to hype up scientific inquiry because, sadly, it needs it. Though you'd think revealing a universe far more intricate and stunning than any ancient script could conjure would be appealing enough. In any case, there is a right way and a wrong way to do it.

Traditional creation myths place humans at the center of the story. This has obvious appeal if you know anything about humans and what makes a compelling narrative for them. While such stories are clearly wrong in a very objective sense, modern interpretations often attempt to distill only symbolic or metaphorical meaning. While this has its own utility, it no longer provides answers to the original question. In its place, the scientific origin story is often understood to be taken literally as an actual history of cosmic events. But this creates a new problem...which is actually the same old problem.

The rise of *scientism*—the belief in science as the ultimate authority—mirrors the rigidity often found in religious doctrines, complete with new dogma pontificated by science celebrities. Replacing liturgical vestments with lab coats, this viewpoint becomes a new form of orthodoxy, where followers of scientism hold scientific hypotheses with unwavering conviction. This is

that surprising since we call many of our scientific theo-
laws of the universe." At its extreme, where science merges
with technology, scientism restores humanity's cosmic celebrity
status—a comforting feeling, no doubt. In our hubris, rather than
simply inspiring a sense of awe and wonder, we imagine this epic
tale of cataclysmic birth and rebirth of the universe, galaxies, and
stars as just our entrance music. We're the main event, and with
our ever-growing technological prowess, the cosmos is ours for
the taking.

Humility

The problem with scientism is that it can become as rigid as the
very religious doctrines it aims to supersede, forming its own
set of unquestionable myths. This breed of scientific dogma-
tism is as zealously defended as any medieval theocracy, with its
acolytes—skeptics, new atheists, and techno-optimists alike—
often engaging in the same kind of "holy wars" as their religious
counterparts. They mount their high horses, armed with data,
charts, and videos of robot dogs, ready to joust with anyone who
dares question the sanctity of science and its final technological
decrees.

In this fight between science and religion, the battlegrounds
range from classrooms and courtrooms to social media and dinner
tables. Each side often paints the other as either hopelessly naive
or dangerously misguided. Whether they remain civil or not,
debates are as pointless as watching children attempting to play
chess with Monopoly tokens. At best, they get hungry and give
up, but it's more likely to end up in kicking, shouting, and crying
because no one can agree on the rules. Holders of creation myths,

such as young Earth creationists, never asked to have their beliefs questioned, nor is it the case that scientists developed their theories for the purpose of disproving individual creationist claims.

Meanwhile, philosophers of science and theologians do engage in good-faith academic discussions. The problem with most of these debates is that they focus on specific questions, such as the existence of God or miracles or, indeed, creation. This gives the illusion that science is a collection of mostly unrelated alternative facts, as if each issue has stand-alone merit. But they don't. To talk about whether capital-G God recently created the world, for example, presupposes a Christian framework, and, unfortunately, addressing individual questions like this indirectly legitimizes it. "Well, your specific science argument might prove the Earth is more than six thousand years old, but it doesn't prove God didn't create it earlier!"

Instead, we should understand how science forms a coherent, comprehensive framework for organizing our knowledge of the environment we inhabit, from the microscopic to the cosmic. Scientists are guided by evidence in a search for truth, not by the hope of negating cultural narratives, even if they end up doing that as an unintended consequence. We trust science not because it contains this fact or that but because its validated predictions are built upon a robust foundation honed through centuries of refinement.

This perspective sees science not as a collection of immutable facts but as a constantly evolving body of knowledge, always open to revision. It's an acceptance that science is perpetually "wrong" in some sense, always on the cusp of being outdated by the next discovery. But why then accept this false narrative over another?

To put it bluntly, we accept the tentative stories of science because they are far more useful than the alternative. Creation myths are deeply rooted in cultural and historical contexts. They offer narratives that reflect the values and understandings of the times and societies in which they were conceived. We can imagine science in the same light, a reflection not of the values of a particular tribe or religious ideology but of humanity. Our global community values progress through knowledge and understanding for the betterment of all life. Science answers this call by providing a common language and methodology for discovering pathways to realize our shared potential to transcend the boundaries defined by our natural individual capacities.

Scientific knowledge is cumulative and self-correcting. It builds upon the discoveries of the past, constantly refining and sometimes revising our understanding based on new evidence. It is our shared heritage, a testament to the collective intellect and curiosity of humanity, a unifying force that transcends individual people, borders, and cultures, bringing us together in a common quest that has already showcased our ability to overcome countless limitations and expand our horizons to places we would never have dared to dream.

Science has humbly removed us from center stage in the universe's grand narrative, as our lives are but a blip in the vast expanse of cosmic time. Yet, they are filled with adventures, struggles, and triumphs. So, while the universe might not revolve around us, our personal universes—filled with loved ones, passions, challenges, and dreams—certainly do. In a final ironic twist, with every discovery, invention, and act of kindness, science reminds us that we are still the heroes of humanity's story. And while every story has its villains—be they the relentless march of time, the existential threats of the cosmos, that asshole down on King Street with the sign telling me I'm going to hell, or that neighbor who insists on mowing their lawn at 6:00 a.m. on a Saturday—these antagonists only add depth to our tale, pushing us to grow, adapt, and perhaps one day write a new chapter where lawns mow themselves and that guy down on King Street reads a second book...perhaps, this one.

2

Stars and planets don't give a shit about you

There is no better way to start your day, or a book chapter, than to consult a horoscope delivered by an oracle who will interpret the divine machinations of the heavens, guiding your easily predictable day of mundane failures, including grammatical errors and run-on sentences that are sure to annoy your editor. And who would this soothsayer be? I asked Google, and apparently, it's the *New York Post*. Well, fuck. This isn't getting off to a good start, is it?

Even if you aren't the last person alive who still reads the *New York Post*, you might still be the type to subscribe to something that gives you your daily horoscope. It might look something like this.

The stars have aligned to provide you with a unique opportunity. Keep your eyes open, and don't be afraid to seize it when it comes your way. The universe is conspiring in your favor.

Sounds good, right? Well, sorry to say, it's complete bullshit. I just made it up, which was tough to do without throwing up in my mouth a little bit. In truth, the only thing the universe is conspiring to do is eventually kill us. But even though the stars have nothing to do with how my day will play out, they are entirely the cause of the phenomenon that is horoscopes.

I'm assuming you know what stars are. If you happen to be out at night and not looking at a glowing rectangular screen, you'll likely see a few of them as specks of light above you. You probably live in or near a city, so you might even be able to count the number of these twinkling dots on your fingers. In other words, the modern night sky is decidedly dull, which is probably why you aren't looking at it now.

Imagine you were alive a thousand years ago, though. First of all, you'd be thanking your lucky stars that you didn't die a horrible medieval death long before you learned to read this. On a happier note—sort of—you'd be treated for half your life to a night sky filled with thousands of stars and the cosmic dust of your own Milky Way galaxy. Of course, you wouldn't know what that was, but it would have been beautiful. You'd stare endlessly at it because you'd probably have nothing better to do in the evenings when it is your turn to be on the lookout for thieves, spies, marauding bandits, escaping prisoners, or that intern who keeps stealing office supplies.

You'd gaze in awe at the mesmerizing spectacle above you, transfixed by the celestial ballet, when suddenly, some self-proclaimed expert, dressed in clothes way fancier than you, would saunter up and impose their interpretations of the heavens upon your innocent stargazing. You'd politely ask, "Who the fuck are you?" And they would overconfidently reply, "I am an astrologer!" And that's about when you'd crunch down on that cyanide pill you wisely took back in time with you.

Connect the dots

Back in the days of yore, people weren't reading horoscopes in the *New York Post*, but they still wanted to know what the point was of all those lights in the sky. Without an understanding of what the stars and planets actually were, they resorted to divine significance. Stare at them long enough, and as with clouds, you'll start to see patterns. We don't know what early humans called them, but for the last six or seven hundred years, patterns in stars have been referred to as *constellations*. Today, the very official-sounding International Astronomical Union recognizes eighty-eight constellations, including old favorites such as Aquarius the water-bearer and Leo the lion as well as more obscure newcomers like Norma the carpenter's level and Puppis the poop deck, which, regardless of the fact that you are laughing, is not a joke. If you know anything about humans, though, the most surprising thing is that there isn't a penis constellation.

In 1928, the entire night sky, splendiferously called the *celestial sphere*, was divided up into these eighty-eight regions so that every observed point of light could be approximately located. Of course, given that the International Astronomical Union is based in Europe, Americans refuse to conform and continue to refer to their own favorite groups of stars like the *Big Dipper*, which is the ass and tail of the great bear, Ursa Major, by the way. The eighty-eight officially recognized constellations consist of forty-two animals, eight humans doing various jobs, nine mythical creatures, and twenty-nine inanimate objects, none of which are penises. It's like an intergalactic zoo combined with a cosmic yard sale—a smorgasbord of random shit with no unifying theme.

An actual photograph of a constellation should look like a

generic photo of the night sky—just a bunch of randomly scattered stars. But, if you google it, you'll find those same stars annotated with line art that looks like an innumerate drunk person's attempt at connect the dots. Ursa Major looks as much like a bear as a photocopy of my ass looks like a Monet—passable, but still a stretch. If you hit the reset button on humanity (not a bad idea at this point), all the stars would be in the same place, but humans probably would have connected the dots in completely different ways. Maybe we'd finally get that celestial phallus we most assuredly deserve.

For the majority of human history, we have seen the sky unaided by the technology we currently take for granted. On a clear night, far from any earthly sources of light, you can see the following with your naked eye: a few thousand stars that appear in a fixed pattern; the five planets you know as Mercury, Venus, Mars, Jupiter, and Saturn, which appear as brighter "wandering stars"; the Moon, 'nuff said; the Milky Way, which appears as glowing dust with bright and dark patches; several small meteors that last a few seconds as streaks of light in the sky; the occasional comet lasting for days or weeks; and very rarely a supernova, which would appear as the "birth" of a new star, even though it's actually the death of an old star that was too far away to have been seen before it booped itself. Oh, and the Sun, which doesn't show up at night for obvious reasons. All this provided plenty of fodder for making familiar patterns of the lights in the sky. But, surprise, the first connect-the-dots books of the stars were actually *audio*books.

Connect the dots, but with words

The oldest written story is four thousand years old—the epic poem about Gilgamesh, a Sumerian king with a penchant for putting mythical beasts in headlocks. Amusingly, this master-piece of ancient literature is also the oldest known instance of plagiarism resurfacing in many forms over thousands of years without anyone ever receiving a phone call from the school. It should be noted that these stories aren't in, like, old dusty books found on library shelves. They were discovered on busted-up and cracked stone tablets. You'd think writing things down would be a foolproof method for preserving stories for future genera-tions, but it seems that scribbling on clay tablets and papyrus isn't exactly the most durable form of recordkeeping. If nothing else, take this as a cautionary tale to keep backups of your stone carvings in the cloud.

Meanwhile, Indigenous Australian stories have been passed down orally, with remarkable accuracy, for tens of thousands of years. No ink, no paper, just good old-fashioned storytelling around the campfire. And yet, these tales have managed to stand the test of time, outlasting entire civilizations and their written records. Who would've thought that the most reliable method for preserving our collective wisdom would be to simply share it with one another and not just rely on hastily scrawled notes stuffed in a drawer somewhere? I only wish they told stories about where I left my fucking keys.

We often associate history with symbols painted on cave walls, etched into stone, or written on paper, but our oldest recorded histories are oral traditions. Some of these include detailed astro-nomical observations. One of the oldest, now preserved for-ever through the proliferation of copies of it on the internet, is "The Emu in the Sky." While not a constellation per se, the name refers to the dark clouds of the Milky Way, which roughly outline the shape of an emu, a large flightless bird obviously native to Australia and also the victor of the Great Emu War of 1932, having defeated the Royal Australian Artillery in a month-long campaign to steal wheat crops. True story.

While myths and cosmological origin stories appear naive through the lens of modern science, they are, in fact, surprisingly sophisticated and often rooted in generations of keen observations of the night sky. Embedding stories in the night sky is ingenious because the star patterns are the only thing that remains fixed for long periods of time. These stories not only serve as a means for imparting moral values but also provide practical knowledge for navigation, weather patterns, agriculture, and understanding the passing of time. The Pleiades, or the *Seven Sisters,* is a particularly striking example of this in the Southern Hemisphere. The Pleiades are a group of stars and the most obvious cluster of bright dots of light that can be seen from everywhere but Antarctica. They can be found in the northwest part of the Taurus constellation.

The Yamatji people of western Australia call the cluster Nyarluwarri. Every year, in late April, it appears close to the horizon as the sun sets. This is the best time to harvest emu eggs for food. The Māori of New Zealand mark the new year when the Pleiades (Matariki in their native language) are first seen at dawn. Descendants of the Inca Empire in Peru call the stars Qolqa, and when the stars only briefly appear at twilight, the people know it is time to plant potatoes. These are just a few examples, while many others have yet to be rediscovered or have simply been erased by smallpox or musket balls. Sigh...there really were "good old days." It's just that they were not fifty years ago, but several thousand.

When humans get together in large communities, they have a tendency to create *culture*. Societies are complex networks that give birth to odd systems of stories, laws, habits, internet memes, communicable diseases, and so on. New things are thus forcefully interpreted within the existing cultural context. Pretty quickly, the system gets so complicated that no single person can comprehend all of it, and we end up with experts who need to be consulted when the need for specialization arises. The problem is that there is no law of the universe external to human affairs that prevents some asshole from declaring themself an expert on the connection between the stars and crop yields or how many virgins you should keep in your harem. Enter, the astrologer.

Why we can't have nice things

The most charitable definition of astrology you can find is *the search for meaning in the sky*. I suppose that would make us all astrologers, though, and thus that's not a very useful definition. It's like saying a thief is someone who searches for things that are valuable to them. Generally, the *practice* of astrology is attributing direct influences of celestial objects on earthly matters. Historically, these included natural things like geological phenomena and seasonal weather. Gradually, it included cultural phenomena, like the politics and laws of cities and nations. Nowadays, it is almost exclusively reserved for divinations concerning individuals, including everything from their innate disposition to their fate in life. How quickly we went from mnemonic campfire stories, through heavenly interpretations of the mundane, to batshit crazy fortune-telling.

To be fair to long-dead people, the astrologer wasn't always bad. To be a "good" astrologer in ancient times meant you had to be a good *astronomer* as well. Astronomy is science—it is the systematic study of celestial objects striving for the most accurate explanations of their nature, including their origin and evolution. Ptolemy was probably the most famous astrologer/astronomer and emphasized that the proper application of astrological principles required the skillful application of measuring instruments and mathematics. Astrology was considered a scholarly pursuit, studied alongside astronomy, meteorology, and even medicine. Nowadays, you can still study astrology, but you no longer require nor obtain any technical skills or knowledge of astronomy. You mostly just have to memorize phrases like, "Your Moon is in the Seventh House of Aquarius, so you'll experience an influx of cosmic vibrations that will enhance your aura's vibrancy," or some shit.

While many forms of astrology exist, the most egregious cosmic-astrology bullshit is based on popular versions of Western astrology. The simplest form is Sun sign astrology. Most people in the West know their "sign." While technically, it has to do with the position of the Sun in the sky at birth, no one actually looks at the sky anymore to determine that. There are simple calendars to figure it out now, or just ask Siri since it knows your birthday. My sign is Aquarius, for example. This is the only information I need to find out which newspaper horoscope applies to me.

With the rise of the mass media and the internet, it has become easier to obtain more sophisticated "charts" that are created based on the locations of not just the Sun but the Moon and all the planets on my birthday. There are even finer details one can pretend are relevant by providing the exact location and time of my birth. Most modern astrologers also include the two planets not visible to the naked eye, Neptune and Uranus, as well other solar system objects such as Pluto and Chiron. Beyond this is pure numerology, which is only connected to the historical roots of astrology through its reliance on superstition.

The primary words used in astrology are based on Latin names given to us by the Romans, who took them from the Greeks, renaming their gods and constellations into what we still call the *zodiac*. The zodiac consists of twelve of the eighty-eight aforementioned constellations. They are Aries, Taurus, Gemini, Cancer, Leo, Virgo, Libra, Scorpio, Sagittarius, Capricorn, Aquarius, and Pisces. Starting with Aries is arbitrary, but this is the order in which they would appear on the horizon as the Earth turns. With a great deal of imagination, they are meant to look like a ram, a bull, a pair of twins, a crab, a lion, a young woman, balancing scales, a scorpion, a centaur, a goat with the tail of a fish, a man pouring water, and a pair of fish. Each of these represents some particular Greek or Roman legend, but there is no coherent story connecting all of them, and modern horoscopes make no reference to the old legends.

This just in

My sneaking suspicion is that if you bought this book, you probably don't know what a "real" horoscope actually is, being smart enough to avoid them altogether. So, what do these stupid things look like? An ancient horoscope would have made reference to the mythology and generically applied it to the city, state, or empire. If they had the *New York Post*...

Salutations, citizens of Rome! As Jupiter, the king of gods, takes his place in the house of Capricorn, a time of expansion and conquest is at hand. Mars, the god of war, stands beside him, promising victories in distant lands. Yet, Venus, the goddess of love, hides in the shadows of the Scorpion, foretelling potential discord in matters of the heart. Offer prayers to Mercury, the messenger god, for guidance in negotiations and clarity in communication to ensure harmony in all aspects of life.

Whereas, now horoscopes are couched in vague, generic terms that could apply to anyone, often relying on the power of suggestion. A modern Sun sign horoscope might look something like this:

Hello, Aquarius! As the Sun shines brightly in your sign, you can expect an energizing day filled with opportunities for personal growth. However, beware of misunderstandings that may arise in your relationships. Remember to keep an open mind and communicate clearly with those around you. Good fortune may be just around the corner, but only if you seize the moment and trust your instincts.

If I ended up going down the rabbit hole, things would get much more convoluted. The horoscope would still be so vague as to apply to anyone, but it would contain the whole gamut of astrological nonsense.

Born under the sign of Aquarius, with your Moon in Cancer

and your ascendant in Virgo, you are an innovative thinker with a strong emotional side and a keen eye for detail. Your Sun-Mars conjunction in the 6th house indicates a strong drive to succeed in your chosen profession, while the trine between Venus and Neptune in your 7th house suggests a deep capacity for love and a romantic, idealistic view of relationships.

Of course, an astrological chart is just a summary of where all the non-stars are at a particular time and location. At some point in history, some enterprising astrological huckster figured out that you can make even more money by telling people where and when they should schedule important events based on astrological bullshit. Speaking of shit...

Ah, seeker of intestinal wisdom, the cosmos has spoken! Your next bowel movement is destined for greatness at precisely 7:42 a.m. on the third day hence. With the Moon in steady Taurus and Mars supporting your health, prepare for an easy victory on the porcelain throne. But beware of Neptune's retrograde—consume high-fiber foods and water to maintain harmony in your digestive realm. Heed the stars, and embrace this legendary event!

I'm not sure if I can hold it. But far be it from me to beat the dead horse of ragging on horoscopes. They have been widely criticized for their lack of specificity and scientific basis and reliance on psychological tricks. In fact, scientists have even given a name for the tendency of people to accept as true information so vague as to be considered worthless—the "Barnum effect." Horoscopes are popular simply because people can't help but find personal meaning in ambiguous statements about their love life and financial situation, the prospects of which are definitely great if only you heed the warnings you have to pay me for!

The good, the bad, and the ugly

Independent of the annoying vagaries spouted by astrologers that a computer chatbot can easily mimic, there are far more insidious misuses of astrological nonsense. First up, *astrological birth control*. No, that's not a joke—there are more books on this subject than I care to count. But they all pay tribute to one asshole, Dr. Eugen Jonas. Jonas is a Czech psychiatrist who, as a devoted Christian, wanted to prevent abortions. To do so, he invented astrological contraception, which he—apparently without a hint of irony—dedicated to the Virgin Mary.

According to Jonas, his method is 98.6 percent accurate at determining which days a woman will be fertile and whether the conception will be a boy or a girl. Thus, and this is 100 percent bullshit, seemingly infertile women can not only conceive children but also choose the sex of their baby. But wait! There's more. Jonas clearly spent some time studying psychiatry since he knows that to really sell products to parents, you have to tap into their fear and anxiety. So, Jonas claims his method can also determine whether the child will have birth defects. The cost of each calculation is about $30.

Now, you might be thinking that fertility is based on a woman's ovulation cycle, but that's too obvious and scientific. Instead— and, again, this is completely false—a woman has a *second* fertility cycle that peaks during the same phase of the moon that occurred at her own birth. The zodiac sign of the Moon at the moment of conception will determine the gender of the baby, and the position of the rest of the planets signals whether there will be complications.

The problem, for Jonas, is that some of these claims can actually be tested, and indeed they have been. While Jonas has never published a peer-reviewed study of the numbers he pulled from his ass, actual scientists have reported, on numerous occasions over many decades, no correlations between birth-related statistics in humans (and also cows for some reason) and the goings-on of the heavens. For due diligence, though, I checked another website where you can pay money for astrologically inspired sexytime recommendations. At least those at thegenderexperts.com are honest with a clear disclaimer:

> Our website and services are intended for entertainment and novelty purposes only. We do not claim a set accuracy rate. We are not a substitute for medical advice, treatment, or diagnosis.

While entertaining, it's not very reassuring since astrology grifting seems to be a slippery slope. In one high-profile case, a French psychic (also a bullshit profession) named Maria Duval was the face of a decade-long mail scam that more wisely preyed on the elderly because it is much easier to manipulate the paranoia of a dementia sufferer than it is to get money off expecting parents who are already about to throw their life's savings into a genetic insurance plan. With Duval's celebrity, an apparently complex network of con artists orchestrated by a man named Patrice Runner, began with the simple act of selling astrological charts and trinkets associated with the paranormal and alternative medicine.

But shipping things is expensive. Why not just ask for money in exchange for empty promises from sick and vulnerable people? While the individual transactions averaged about $40, the scam added up to more than $175 million in the United States alone, and the government found it difficult to shut down. So it turns out the stars can influence earthly matters after all, but only if it's mediated by petty swindling.

Adolf Hitler. Oops. I should have warned you first that he was going to show up. As you know, Hitler had many shitty ideas, but he also allegedly believed horoscopes to be nonsense. However, that didn't stop Louis de Wohl, a Hungarian author, screenwriter, and con man who fled Germany for the United Kingdom in 1935, from convincing British intelligence of the opposite. De Wohl claimed that Hitler was very superstitious and relied on astrology to make key strategic decisions. The next step in his argument is actually quite logical—if the British Secret Service had accurate astrological charts for Hitler, they would know the exact same information Hitler received from his fortune tellers. Luckily, de Wohl could produce these. So, at the height of the deadliest conflict in human history, a fake astrologer was determining the likely dates of Nazi campaigns based on what the sky looked like at the time and place of Hitler's birth. Yeah...

On a lighter note, Ronald Reagan. Ah, now that's better. It's nicer to picture a famous movie star, the president of the Screen Actors Guild and also the United States. Another claimed world leader with irrational superstitions? It can't be. But this time, oh yes, it can. In his memoir, the chief of staff for President Reagan wrote:

> *Virtually every major move and decision the Reagans made during my time as White House chief of staff was cleared in advance with a woman in San Francisco who drew up horoscopes to make certain that the planets were in a favorable alignment for the enterprise.*

Holy shit.

The Great Randall

By the time the Scientific Revolution culminated in the publication of Isaac Newton's laws of motion in the late seventeenth century, astrology was all but dead as an intellectual pursuit. Luckily for practicing astrologers, the general public has never really paid much attention to the tiny fraction of humanity labeled as academics. So astrology not only lived on but also thrived in the cesspool that libertarians love to call *the marketplace of ideas.* But the tools of science also evolved, and after two hundred years, astrology was due for its second reckoning.

When people claim that astrology has been "debunked" and label it *pseudoscience,* they often do so within the very particular context of experimental testing. Karl Popper, an intellectual hero of nerds and science bros alike, famously equated science with *falsification*—that a theory is scientific only if it can be proven false through empirical observation. For Popper, this was crucial in demarcating the line between actual science and bullshit, which he euphemized as *pseudo*science. His favorite example? Astrology.

It has the "flavor" of science in that it claims to rely on evidence, but it only includes favorable anecdotal evidence and is not open to falsification. In doing so, it enjoys the company of relative new-comers to the pseudoscience party, such as ancient aliens, moon landing denialism, the hollow Earth, the flat Earth, the young Earth, aromatherapy, ghosts, and thousands more.

Thomas Kuhn, Popper's younger contemporary, famously disagreed, arguing that astrology shouldn't even be considered pseudoscience since, for the most part, it doesn't purport to follow the scientific method at all. Instead, astrologers operate within their belief systems separate from, and sometimes blatantly anti-thetical to, the scientific paradigm. The debate was never settled. But there was another movement emerging around the same time that would take the fight against astrology in a different direction. Enter the skeptic community. They couldn't be bothered with nuance, so they barreled full steam ahead on the Popper train.

For simplicity, imagine there are only four groups of people in the world: the evil charlatans, the hopelessly naive majority, the dispassionate scientists, and the stoically rational skeptics. In this story, the charlatans obviously prey upon the naive. They don't need to be clever about it either. They just need to know where to steal tried-and-true methods of deception from. (It's the internet, by the way.) Using the tools of science, the scientists can clearly see through the ruse, but they don't engage because that would take time and money. The skeptics, who seem to have a lot of the former on their hands, come to the rescue, slaying the evil char-latans and saving the naive people of the world, who laud them as heroes as everyone lives happily ever after—except the scientists. They still have work to do.

Of course, that's only a fairy tale that science bros like to believe. In reality, the charlatans aren't evil. They are actually hopeless, which is what drives them to fraud. The majority aren't naive; they just don't care. And the skeptics? They are mostly just podcasting. The scientists are the same, though. They really don't have the time or money to devote to studying pseudoscience. This is not an excuse or justification for elitism. Scientific studies are costly and take a lot of time. Would you rather have scientists performing multidecade longitudinal studies about cancer treatments or the efficacy of astrological charts? Because you can't have both. Skeptics, on the other hand, actually study pseudoscience, not because they are masochists but precisely because they have a strong belief that purveyors of astrology need to be stopped. In the words of Steven Novella, one of today's most prominent skeptics:

> Skeptics endeavor to protect themselves and others from fraud and deception by exposing fraud and educating the public and policy-makers to recognize deceptive or misleading claims or practices.

Armed with the tools of science, this group of thinkers, writers, and entertainers seek to expose and debunk the myriad of pseudoscientific beliefs that have taken hold in popular culture.

The first professional skeptic was probably James Randi, also known as The Great Randall or The Amazing Randi. A successful magician by trade, he became a prominent figure in the skepticism movement, using his knowledge of trickery to expose pseudoscience frauds. He became quite famous, appearing on multiple popular American TV shows at a time when being popular meant you appeared on popular American TV shows.

As Randi and others took on astrology, they encountered a problem similar to the one Popper and Kuhn faced. Namely, how do you effectively debunk a belief system that doesn't necessarily operate within the bounds of science? The approach taken in public appearances was mostly mockery and ridicule. While amusing for many, it wasn't always successful in changing minds. In fact, it occasionally had the opposite effect, inadvertently emboldening existing believers and even creating others.

Randi was thrust into primetime stardom through *The Tonight Show* with Johnny Carson, who was apparently a fan. As luck would have it, shortly after the Carson-Randi bromance began, *The Tonight Show* booked Uri Geller, a self-styled psychic and the butt end of previous public shaming by Randi. Carson consulted with Randi, who advised the show on a particular set of props that would evade Geller's standard tricks. The result was epic humiliation—Geller assumed his career was over. On the contrary, though, he was immediately booked on *The Merv Griffin Show*, where he bent some spoons with his mind and became a celebrity.

Mentioning television shows that can't later be streamed on YouTube might seem like ancient history to some, but Geller is still going strong. However, the exercise in narcissism he calls his website might suggest otherwise. Here is where you can buy citizenship to his private island—called Mystical Island—for $1, and also where it becomes apparent that he has an obsession rivaling Salad Fingers with bent spoons. Geller displays all his junk as artifacts in a renovated soap factory he calls the Uri Geller Museum. The tour guide, Uri Geller, will show you a Cadillac with 2,000 bent spoons riveted to it as well as the largest spoon in the world, a 16-meter-long sculpture of—you guessed it—a bent spoon. And, if you ask him to bend a spoon, he will show you a tattoo of a spoon on his arm that "bends" with his elbow.

Now, you may be wondering why I am telling you about what appears to be a much lamer version of *Tiger King*. The point is that if the only options for debunking pseudoscience are expensive and time-consuming scientific studies or derisive public takedowns, we are essentially screwed. The only effect will be in providing the attention that the offender either craves or requires. Indeed, even impartial sources of information will claim that astrology has no scientific validity. Yet, it is as popular as ever.

Debunking debunking

To anyone with at least a little bit of scientific literacy, astrology is so comically flawed that it certainly seems it would be easy to refute in some formal and rigorous way that would satisfy any rational person. But, clearly, it isn't that easy. So who do we blame for that? Since he is now dead and can't defend himself, I suggest we blame Popper. The problem with the popular version of the "scientific method" inspired by Popper is that it is alluringly simple. The basic formula is as follows.

1. **Observation:** Make a careful and systematic observation of the natural world.
2. **Question:** Ask a question about the observed phenomena.
3. **Hypothesis:** Formulate a testable hypothesis that provides an answer to the question.

4. **Experiment:** Design and perform experiments to test the hypothesis.
5. **Conclusion:** Draw a conclusion based on the analysis of the experimental data.

It is so simple that one can say both multibillion-dollar particle accelerators and mock volcanoes filled with baking soda are equivalent applications of science. Imagine if you are a judge at the local high school science fair and you walk up to a submission clearly made by the overachieving offspring of helicopter parents who have spent sleepless nights perfecting their precious progeny's project.

1. **Observation:** Mixing baking soda and vinegar produces a fizzing reaction.
2. **Question:** What causes the fizzing reaction between baking soda and vinegar, and how does the reaction work?
3. **Hypothesis:** The fizzing reaction is caused by a chemical reaction that makes gas bubbles.
4. **Experiment:** I carefully mixed different amounts of baking soda and vinegar in a container and observed the resulting reaction.
5. **Conclusion:** Based on my observations, the hypothesis about chemical reactions making gas bubbles is supported.

That's a solid C+, Timmy. Next up...wait, CERN? How did they get in here?

1. **Observation:** Scientists have observed that the universe is made up of particles and forces that interact with one another.

2. **Question:** How do subatomic particles interact with each other, and what new particles might exist beyond the known particles?

3. **Hypothesis:** A new particle we call the *Higgs boson* exists and is responsible for giving other particles mass.

4. **Experiment:** We built a particle accelerator the size of a small city and smashed high-energy particles together, collecting data from six quadrillion collisions with detectors half the length of a football field.

5. **Conclusion:** Based on forty years of experiment, data, and analysis, we conclude that the Higgs boson has likely been detected.

Also, C+, show-offs. Both of these experiments follow the same "scientific method." So we seem forced to conclude they are equally valid applications of science. Yet, clearly thinking about science as only a method to test hypotheses blurs the lines between backyard experiments and cutting-edge research. Popper's approach puts undue importance on only hypothesis testing, experimentation, and evidence-based conclusions, which leaves the term *science* open to abuse, which we can see copious evidence of when it's applied to astrology.

1. **Observation:** Since the time of ancient civilizations, humans have noticed that celestial patterns seem to line up with events on Earth, including the personal lives of individuals.

2. **Question:** Can the positions of celestial bodies, such as stars and planets, influence human behavior and the course of events on Earth?

3. **Hypothesis:** The positions of celestial bodies at the time of a person's birth can determine their personality traits, relationships, and life events.

4. **Experiment:** Ummm…track the positions of celestial bodies and compare them to people's lives, looking for patterns and correlations?

And this is where the scientific method either remains at the station or goes off the rails. In something that sounds way too charitable but is unfortunately true, we must say that science—or at least Popper's vision of it—is inconclusive about astrology, and that ambiguity has allowed it to persist.

The problem with applying the scientific method to astrology is that any hypothesis you might come up with is impossible to test in a rigorous and controlled manner. If you do manage to come up with one, as many researchers have, it will necessarily be too narrow for any astrology supporter to accept. They can simply claim the hypothesis is not relevant to *their* version of astrology.

In most cases, you can't really make convincing "evidence-based" conclusions because people want yes-or-no answers, and there is never enough evidence for certainty. The astrologer can always claim gaps are enough room for doubt. Worse, you can easily appear *as if* you are applying the scientific method with plenty of evidence. Meanwhile, you are drawing conclusions from corrupted data or improper analysis, which often requires someone trained in statistics to perform.

Without the ability to actually apply science to questions surrounding astrology, debunkers often resort to humiliation stunts targeted at the least sophisticated examples of pseudoscience. In most cases, these are loosely interpreted as applications of the scientific method. If skeptic conventions had science fairs, you might see something like this.

1. **Observation:** People use language from astrology to sell bullshit horoscopes.
2. **Question:** Is there any objective truth to horoscopes?
3. **Hypothesis:** Horoscopes are vague and agreeable to anyone regardless of their date of birth.
4. **Experiment:** I stole a single horoscope from an astrologer, gave that same one to ten people with different Sun signs, and asked them if it was accurate.
5. **Conclusion:** Since eight out of ten people said it was accurate, I conclude that this astrologer is full of sh...er...I mean, I conclude that my hypothesis is correct... Also, the astrologer is full of shit, and all of astrology has now been debunked. Oh, and for good measure, I also told the participants that they were given the same horoscope just so they look like fools too. Ha-ha!

While everyone secretly delights in the misfortune of others, these guilty pleasures become almost euphoric when the "other" is someone you morally object to. But as we laugh along at the humiliation of crooks and the embarrassment of willfully gullible people, the point has somewhere been lost. However, if you feel you might be susceptible to the seeming authority of anything couched in statistical language and want to practice avoiding being embarrassed, I suggest you look at the examples in Tyler Vigen's *Spurious Correlations*, where it is demonstrated that near-perfect correlations over an entire decade exist between things like the number of films Nicolas Cage has appeared in and the number of people who drown by falling into a pool. Actually, that might not be a coincidence.

Mental models to navigate the world

It's not Popperian science that renders me, and most other scientists, immune to astrological bullshit. It's actually the entire body of scientific knowledge that contradicts its underlying premises. By shifting our focus from the empirical method to the underlying theories and principles that form the basis of scientific understanding, we can further expose the inconsistencies and implausibilities inherent in astrological beliefs. This transition allows us to delve deeper into the philosophical underpinnings of science, shedding light on the stark contrast between the rigorous explanations provided by modern science and the unfounded assumptions that characterize astrology.

Through its now rich history, science has become a complex framework for understanding the universe and its workings. This framework consists of interconnected theories, principles, and mental models that help us navigate and make sense of the world around us. By continually refining these models through observation, experimentation, and critical analysis, we gain a more accurate and comprehensive understanding of the natural phenomena that shape our existence.

Cosmology is the study of the origins, evolution, and structure of the universe at the grandest scale. While it is undoubtedly the most audacious branch of science, it is still grounded in the fundamental principles of physics, including general relativity and quantum physics. These principles are not simple hypotheses that have been tested once in a science fair experiment; they have been rigorously scrutinized and are the subject of ongoing experiments that happen every moment in government, university, and industrial laboratories all over the world. The conclusions drawn from cosmology are thus incalculably more reliable and consistent than some ideas derived from the anecdotal evidence of magical thinkers.

Let's run through some of the lessons of cosmology that are inconsistent with the underpinnings of astrological thinking. In doing so, it's going to look like ancient people were rather stupid. If you feel that way, keep in mind that, biologically speaking, humans haven't changed for millennia. We are *Homo sapiens*, a species that has existed for over 200,000 years. If we had "evolved" since then, we'd be a different species. You are basically a cave dweller with the immense advantage of a collective intelligence held together by the fragile strings of globalization. Don't pretend like you'd last more than five seconds on *Alone* before you started praying to the stars to go home to your precious Wi-Fi.

What is a star?

Many pseudoscientific ideas appeal to "ancient wisdom," and while some pearls do exist, prescientific concepts of the nature of stars are incredibly naive. Ancient cultures often had their own unique interpretations of what stars were and how they influenced the world. Some believed that stars were divine beings, while others thought they were holes in the sky, allowing light from the heavens to pass through them, like that tent your dad refused to replace and forced you to sleep in during family camping vacations back before iPhones were a thing. Such proclamations were still made in public up until as late as the 1970s when the *war on drugs* effectively halted the recreational use of psilocybin.

For the ancient Greeks, stuff on Earth was made of four elements: fire, air, earth, and water. They believed that stars were made of a divine non-earthly material called *aether*. Later, alchemists thought they could create this fifth element in pure liquid form by distilling alcohol over and over, which, unbeknownst to them, just made pure alcohol. I won't argue with claims that this had transformative power. In modern Hollywood, *The Fifth Element* is apparently composed of starlets willing to undress in front of a confused-looking Bruce Willis while Chris Tucker screams at them. We've clearly come a long way.

Astrologers themselves generally did not focus on the composition of stars. The primary concern of astrology is interpreting only the positions of celestial bodies and their perceived influence on human lives. Today, astrologers are happy to go along with modern scientific concepts, such as the fact that planets are not, in fact, "wandering stars" but balls of rock and gas that orbit the Sun. The geocentric model that was overturned during the Scientific Revolution seems to have been more of an inconvenience for astrology. Astrologers are happy to accept scientific facts that don't directly contradict their beliefs. They might even repeat the fact that stars are burning plasma balls of nuclear explosions held together by gravity just to avoid appearing as antiscience denialists. In any case, the composition of stars never appears in a horoscope.

But knowing what a star actually is contributes strong

indirect evidence against astrological beliefs. To ascribe a fixed and unchanging meaning to stars requires that stars themselves are fixed and unchanging. But stars are not fixed. They are born, evolve, and eventually die over the course of millions to billions of years. During their lifetimes, they undergo various changes in size, temperature, and brightness. Like the addition of previously unknown planets, astrology could be modified to have these facts also tacked on as well. This would result in only adding further complexity to an already complicated framework, whereas science has produced a *simpler* explanation of not only the visible phenomenon but also those discovered with the aid of technology. But, if you really want a more complicated story with no predictive power, by all means, hop on the astrology train to nowhere.

While stars do not give a shit about your life, they do lead lives of their own. Stars are born from vast clouds of interstellar dust made mostly of hydrogen. Gravity, the "force" that exists between everything in the universe through the mutual attraction of their warping of space-time, is relatively weak but eventually created clumps in the cloud. Subtle irregularities cause a swirling effect that creates the rotation of the eventual star and its accompanying planets. When the hydrogen atoms in the center of the clump are compressed to the point of near contact, subatomic forces take over, creating nuclear fusion. Fusion releases energy, which exhibits an outward pressure counteracting gravity. Eventually, an equilibrium is reached, and a stable star is born.

What happens after the birth of a star depends on how big it is. As noted in the last chapter, the Sun is a medium-sized and middle-aged average star, which also pretty well describes your humble author. An actual star is far more predictable than a human, though. Hydrogen is its fuel, and it doesn't make more than what it started with. As a star exhausts its hydrogen fuel, its core contracts and heats up, causing more fusion and the outer layers to expand. The star becomes a red giant, with its outer surface significantly larger and cooler than before. Eventually, the helium in the core starts to fuse into heavier elements. When the helium fuel is exhausted, the process repeats until pulsating contractions and expansions give way to a final ejection of the star's outer layers engulfing anything nearby—which would include Earth for our star. From a great distance, though, it looks like a beautiful puff of cosmic gas and dust we call a *nebula*. The remaining core becomes a white dwarf star, which will cool and fade slowly until the end of the universe. For the Sun, which was born under the star sign "fuck you," this has been fated to happen in about five billion years. No human will be there to interpret the meaning of it.

In contrast, massive stars live that rock-star lifestyle, burning up their fuel much faster and experiencing a much more violent end. When they exhaust their nuclear fuel, the core collapses in a matter of seconds, causing a powerful explosion known as a supernova. This event releases a tremendous amount of energy and can briefly outshine an entire galaxy. The remnants of the explosion of a star that begins with about three times the mass of the Sun is a neutron star. If the star began with roughly ten times more hydrogen than our Sun, the result is a black hole.

Black holes are the most enigmatic objects in the universe. They have such an intense gravitational pull that not even light can escape, rendering them invisible to direct observation. Yet, with the tools of science, we can detect their presence through their influence on the objects around them. When two black holes merge, they send out powerful gravitational waves that ripple across the fabric of space-time. Repeatedly doing so eventually results in objects we call *supermassive* black holes. When these merge, we can now detect them with giant laser-based antennas so large that they must take the curvature of the Earth into account to keep them straight. This has opened up a whole new way to study the universe that makes even modern telescopes look antiquated.

My favorite black hole is Sagittarius A*, located at the center of our Milky Way galaxy. It is roughly four million times more massive than our Sun (read that again). Naturally, it is located in the Sagittarius constellation, though it is not one of the visible stars for the previously stated reason that it is an enormous light and matter-eating leviathan. This really makes you question what it means to be "in" a constellation. For if you knew what was there, you'd definitely not want to visit. At least, your horoscope ought to be more reflective of the facts.

Sagittarius. This week, the supermassive black hole Sagittarius A* at the center of our galaxy looms large, inviting you to ponder the void and embrace existential dread. As the gravitational force of this cosmic enigma mirrors the burden of your own thoughts, your financial and romantic prospects wane in the face of the overwhelming darkness. Find solace in the fleeting nature of existence and the transience of even the most luminous celestial bodies. Embrace the uncertainty of life, knowing that in the grand scheme of the universe, nothing truly matters. Let the darkness of Sagittarius A* guide you to a deeper understanding of the human condition and the beautiful uncertainty of existence.

All four fundamental forces are at play in the ever-changing life of a star. Clearly, if we are going to anthropomorphize stars at all, we should probably say they have their own shit to deal with, which is infinitely more pressing and awesome than the tribulations of little ants stuck on a mostly inhospitable planet revolving around an otherwise unremarkable star who don't do much other than continually make poor financial decisions. The constant transformation of stars challenges the notion that their positions and attributes have fixed, unchanging meanings in astrology. In the end, though, it doesn't matter because the Sun is going to eventually die and, in a final act of spite, take astrology with it.

Where are the stars?

Speaking of Sagittarius, you might wonder why the universe arranged a bunch of stars in such a way as to resemble...what? I don't see it. A centaur archer drawing its bow? Fuck it. I give up. There's really no point in straining to find whatever shape these stars are purported to outline. Those born in early December, armed with a bit of knowledge of the fact that stars are faraway balls of perpetual nuclear explosions, might still ask, "How far away is Sagittarius with these stars that decide my fate?" But this is a silly question because the reality is that the stars that make up Sagittarius are scattered all over our region of the galaxy. Astrologers will rarely point this critical fact out.

Have you ever been to one of those 3D art exhibits that looks like a bunch of random shit except from a very specific point of view where it looks like an elephant or something? That's Sagittarius. It's an optical illusion not dissimilar from that created by every wannabe Instagram influencer with the totally original idea of pretending like they are holding up the Leaning Tower of Pisa. Consider the fact that even the two brightest stars in Sagittarius, called Kaus Australis and Nunki, are about 10 trillion kilometers apart, even though they seem to appear right next to each other in the sky. This means that by the time Nunki has learned of your birth, Kaus Australis will have already witnessed your death.

Nunki is not less bright simply because it is farther away, either. In fact, it is ten times brighter than Kaus Australis, but 10 trillion kilometers is really far. Nunki is also eight times more massive than the Sun, which means when it dies in a cataclysmic supernova, it will leave behind a black hole. Because of its size, it burns through its hydrogen fuel much faster than smaller stars. This means it is so hot it shines blue. It also only started shining about thirty million years ago, long after the last dinosaur charted the stars and the continents last moved. A common ancestor between humans and apes might have looked up to see its birth (with a lag of a few hundred years, of course). Nunki is already halfway through its life, much like the Sun. But whereas the Sun has five billion years more to go, Nunki will be toast in about twenty-five *million*. I'd like to say that the complete disappearance of the archer's arrow ought to put a damper on its divine inter-pretation, but a lot of cosmic shit is going to go down before then.

The third brightest star in Sagittarius is Ascella, which is half the distance between Kaus Australis and us. Something you won't find in your horoscope is the fact that Ascella is not a star at all but *two* stars that orbit one another once every twenty-one years. And, fun fact, Ascella is not a two-star system at all but a *triple-star* system. However, recent observations suggest this third star might not orbit the others and is simply another faint star in the same line of sight. Now, isn't all this way more interesting than finding out the lucky numbers for the lottery weren't so lucky after all?

The vast distances between stars make the whole "constellation" concept seem pretty arbitrary, but that, of course, doesn't stop astrologers. Celebrity astrologer Aliza Kelly wrote in an *Allure* magazine "explainer" to help you understand the importance of the arbitrary pattern these stars make.

> *Represented by the archer (a half-man, half-horse centaur), Sagittarius isn't afraid to use its bow and arrow to explore expansive terrain, seeking answers in places and spaces others wouldn't dare venture.*

Sigh. There is no fucking archer. Drop it. If you were actually standing in front of the Leaning Tower of Pisa, you'd see a single person appearing to hold it up and dozens more looking like idiots, grimacing while pushing air. From any other vantage point than ours, those stars would make a completely different but equally arbitrary shape. There is nothing special about our vantage point on these randomly placed sources of light. There's also nothing special about our "wandering stars" either. Astronomers have discovered more than a hundred planets orbiting other stars in Sagittarius alone, and more are constantly being found.

What's worse is that if alien astrologers existed on any of those planets, they wouldn't even see the feeble light from our own star. How pathetic. It's too bad because I bet we totally would have rocked it as part of a penis-shaped constellation.

The tacit assumption in astrology is that the stars are clusters of cosmic significance, carefully arranged by the universe to guide our lives. But the reality is that they vary vastly in distance, size, and brightness, with no inherent patterns or meaning beyond what we choose to ascribe to them. Many aren't even single stars but multistar systems, and new stars are continuously forming while others fizzle out or explode. Humans delight in finding patterns, and I prefer order to chaos as much as the next person, but I can't imagine basing my life decisions on a cosmic game of musical chairs.

Stellar motion

Stars appear to form a fixed pattern in the sky—an apparent celestial backdrop against which the rest of the action plays out. The North Pole always points at Polaris, the North Star, while the rest rotate around it. Similarly, the South Pole points toward Polaris Australis while the rest rotate around it, but in the opposite direction, just like the toilets in Australia. (Both of those are jokes, by the way.) But the Sun, Moon, and planets do not remain fixed relative to the stars—they move "through" the constellations. However, if you looked carefully at those posters of the solar system in your science class, you'd notice that all the planets lie on a plane. This means that they all follow the same path as the Sun does through the sky called the *ecliptic*. As Earth and the rest of the planets go around the Sun, they appear to "visit" the same sets of stars periodically. The Moon travels through them all once every twenty-seven days, while the Sun does so once per year, which is why Sun signs change throughout the year before repeating.

As you will also recall from science class, or seeing a globe, the Earth rotates on an axis that is tilted about 23 degrees, which is the cause of the seasons, and the reason the Sun doesn't appear to follow the same path throughout the year from a fixed location on Earth. It's higher in the sky during summer and lower during winter (in the Northern Hemisphere). As the length of the day changes, different constellations "rise" at dusk. These are standard science facts that astrologers will even use to give the illusion of authority or legitimacy. The problem is that the periodic spinning of Earth on its axis and around the Sun is nothing compared to its motion relative to the stars.

While traditional astrology relied heavily on the idea that Earth was the stationary center of the universe, modern astrology doesn't seem to be fazed by the fact that the Earth spins around the Sun at a blistering 100,000 kilometers per hour. The conceptual center of the universe is now the Sun, and Earth returns to the same place it started every year. Of course, that's not true either. The Sun rotates around the center of the galaxy at 800,000 kilometers per hour. There are roughly 100 billion stars in the Milky Way. The hundred or so stars that form the basis of astrology are relatively nearby but travel around the galaxy at slightly different speeds. So, even ignoring the birth and death of stars, the sky looks slightly different every night.

Remember Ascella? It's actually moving 80,000 kilometers per hour away from us, probably to prove a point. A million years ago, when our ancestors first stood up to pee, Ascella would have been over ten times brighter than it is now—that's three times brighter than the current brightest star in the night sky (Sirius). Imagine how many stories have been told about the divine significance of a star (which is really two stars) that decided to just fuck off. The stars, and hence the constellations, though not really changing themselves, look significantly different *from Earth* on timescales of about 100,000 years. For example, thousands of years from now, Orion's shield will get smashed, and he'll lose his head. Presumably, astrology will still be around if humans make it that far. If nothing else, they are a creative bunch when it comes to bullshit. I can already see it now...

In this new age of Orion, the astrological implications for those born under this sign have shifted. His once gleaming shield, a symbol of strength and protection, has been shattered by the celestial tides, revealing the vulnerability that lies beneath. The absence of his head speaks to the profound notion that our thoughts and ego can sometimes cloud our perception. No longer the valiant, headstrong warrior, Orion now calls upon its followers to embrace humility, vulnerability, and acceptance of the ever-changing cosmos. Embrace the change, dear Orion-born, and may the wisdom of the headless hunter guide you on your path.

Of clearer consequence is the Southern Cross, which will end up looking like two parallel lines, meaning a bunch of people in Australia with missing teeth and unintentional mullets will have to have their tattoos removed, and several other nations will have to redesign their flags.

The motion of stars significantly impacts the validity of astrology. Astrological signs and interpretations are based on outdated, static views of the cosmos that don't account for the dynamic and ever-changing universe we actually inhabit. This makes the practice of astrology inherently flawed, as it ignores the fundamental principles of cosmology, which are inconsistent with the idea that our personalities, life events, and fate are determined by arbitrary patterns in a snapshot in time of an ever-shifting sky.

Forces of nature

Here's a fact that you should take with a generous helping of salt: the stars *do* have a direct influence on you through at least one of the fundamental forces of nature. The two strongest forces operate only on subatomic scales, so it is safe to say they play no role in mediating any purported influence from distant stars.

Gravity is the most apparent force of nature, even though it's actually the weakest of the four fundamental forces. Yes, the Sun and the Moon have a significant effect on Earth, as evidenced by the tides, but they are really close to us. Some of the stars in the sky are large, but they are also extremely far away. The gravitational pull on you from some of those stars would be weaker than any one of the human-made satellites going over your head right now.

The International Space Station is passing through Taurus! What does it mean?

While that is certainly meant to be a joke, the International Space Station and Kaus Australis do have about the same gravitational effect on you. But there are two things to consider. First, it's stupidly tiny, like less than a millionth of a millionth of the force I'm using to push down the keys on this keyboard right now. Second, and this is probably the most crucial fact of all, the influence of every force in the universe is limited by the speed of light. This is why cosmological distances are measured in light-years, the distance light travels in a single year. Kaus Australis is 143 light-years away, which is why if you pray to it for strength on your next centaur-worthy adventure, it will never get the message in time. Meanwhile, the International Space Station is a thousandth of a light second away from me every ninety minutes. So, sixteen times per day, there is a tiny pull on me from the cosmos equivalent to that of a star. How quaint.

Speaking of light, that is the one force from the stars that directly influences our lives. But pretty much all of the energy available for life to thrive on Earth comes from one star—the Sun. Light is our experience of the electromagnetic force. It travels as waves to mediate electromagnetic interactions. Traveling at the speed of light, naturally, means that it seems instantaneous here on Earth when you flick a switch or place a magnet on the fridge. But the light from the Sun takes eight minutes to reach Earth, so when you look at it—hopefully not at noon—you are seeing it as it was eight minutes ago. The idea that we really experience the past becomes existentially more terrifying when it comes to the stars.

At a single point in history, we could see a star from outside our own galaxy with the naked eye. SN 1885A (or *Supernova 1885*) was seen in, well, 1885. Studied mostly with telescopes—since we had them—it was bright enough to be briefly seen on a clear night without aid. SN 1885A occurred in the Andromeda galaxy, which is well over two million light-years away. In other words, the star died, and we didn't find out for millions of years. Andromeda is the most distant object we can see with the naked eye. It was known to ancient astronomers as a little dust cloud but never made it into the lexicon of popular astrology. Perhaps Andromeda anticipated this and decided to blow up one of its stars to signal its annoyance at being left out. Or, perhaps, it was just serendipity for a few amateur astronomers.

While Andromeda is far away now, our two galaxies are on a collision course, set to eventually merge into a globular-shaped super galaxy we've already called Milkdromeda. While the view from Earth will be spectacular, it will also occur at the same time as our Sun is dying. I hope humans are still around somewhere in Milkdromeda, though, continuing to chart the stars—only this time as travel agents instead of peddlers of bullshit.

3

Aliens were never here, and they aren't coming either

If someone were to knock on your door in the middle of the night, offering you an adventure of a lifetime through interstellar space, complete with the chance of meeting extraterrestrial life, your first logical step would probably be to call the police. On the other hand, you might be more of a free-spirited type, or maybe you really took to heart that empowering message on your coworker's aspirational poster to *seize the moment*. In any case, I've already made the decision for you, so grab that bottle of hand sanitizer—who knows where this spaceship has been—and hop aboard.

Now, let's set some expectations. This won't be a Douglas Adams–esque romp through the cosmos. No, think of it more like a Michael Moore documentary, but just the B-roll footage of him traversing the Midwest through desolation and no signs of intelligent life anywhere. Because as far as the actual intelligent life on Earth can tell, there is nothing out there—no eccentric galactic politicians, no artificially intelligent beings or paranoid robots, not even a speck of mold hidden under a rock. And don't even get me started on how boring space would be.

But even going out there and seeing firsthand the lack of evidence of life elsewhere in the cosmos wouldn't be enough for the true believer obsessed with the idea that gray-skinned, bug-eyed beings are out there, zipping around in their flying saucers, abducting cows, leaving weird patterns in Farmer Bob's cornfield, and providing unsolicited colonoscopies for lonely pub patrons. Whether it's just a stubborn "want to believe" or insatiably curious what-if situation, the true believer will never be dissuaded.

For the modern technology-obsessed nerd, the goofy shenanigans of pop culture aliens are an obvious farce. Instead, they've been sold their alien dream by big-budget sci-fi movies, with Hollywood flashing images of advanced civilizations with a hive mind that, for some reason, is hell-bent on destroying our already fucked planet. Yet, even though most of us are well versed in the tales of *Men in Black* and *Independence Day*, few have any idea about the real, scientifically backed quest to find our supposed neighbors in the sky. The search for extraterrestrial life is as much an academic pursuit as any other question in astronomy. Though, spoiler alert, it doesn't involve Will Smith punching any aliens...or people.

Are we alone? You should already expect the least exciting answer, but science is so drab, isn't it? You're probably still just a tad curious, the *X-Files* theme song crescendoing in the back of your mind as all reason is replaced by a tagline—*The truth is out there*. Fine. Let's do this. Because whether or not we're alone in the universe, the stories, myths, and beliefs that have emerged from our collective imaginations are a testament to our endless curiosity and, perhaps, our underlying hope of finding connection in the vastness of space...because we sure as hell aren't going to find it in Missouri.

Where did aliens come from?

Depending on what generation you are from, you might imagine aliens as either little green men, lanky gray mind readers, machines that transform into American-brand vehicles for some reason, or just whatever James Cameron's latest nightmares were made of. But the first aliens were none of these, and they were infinitely more badass.

Consider Ra, the Egyptian god who carried the Sun across the sky in the ancient equivalent of a spaceship (a.k.a. a boat). Or take Odin, the Norse god who lived on something of an exoplanet called Asgard, connected to Earth by a goddamn space elevator called the Bifröst. And what about the Hindu god Shiva, with powers that not even Tony Stark could imagine protecting Earth from? Let's also not forget that back when the Earth was still flat, deep below the surface was just as much an otherworldly place as the heavens. With sick alien tech like an invisibility helmet, the Greek god Hades kept a three-headed pet and controlled the interplanetary shipping lanes—just don't forget the tip!

To make sense of the vast, unfathomable cosmos, our ancient ancestors needed narratives, and why not add some dramatic twists? Not only did gods, deities, and divine beings war with each other in their imagined realms, but they also occasionally graced us lowly Earth dwellers with their presence. However, being "touched" by an angel was not likely as heavenly as daytime television would have you believe. From Sumerian and Christian myths of demons seducing unwilling humans to Zeus siring countless demigod children with unwilling mortals, molestation seems to be a recurring theme in alien lore.

You might wonder what the connection is between Zeus

duping virgins into sleeping with him and small gray men with godlike technology but decidedly unsophisticated medical interventions. At their core, both of these figures are external entities that have weird sex addictions and powers beyond human comprehension. The only difference is the "somewhere else" has changed from "up there" to "out there."

Science to the rescue?

The Scientific Revolution dramatically shifted our perspective on the cosmos. The Earth turned from a flat island held up by a turtle at the center of a small universe to a lonely sphere barreling around the Sun with the rest of the planets, all lost in an infinite expanse. Unfortunately, with the advent of the printing press and the eventual appetite to distribute more than just Bibles, our imaginations could now go viral along with science.

Today, it seems science is not enough. Reporting on science *must*, the outlet demands, be couched in wild speculation—or else the viewers just aren't going to see enough ads! But sensationalizing science is not new. In one of the first attempts at popularizing science way back in 1686, French writer Bernard le Bovier de Fontenelle expounded upon the core ideas of the Scientific Revolution, including the fact that the Earth is just another planet and that all the stars are Suns themselves, likely illuminating their own planets, which—just to sweeten the deal—are home to aliens.

Fontenelle also speculated on the nature of inhabitants of our own Sun's planets. For example, Venusians are "a little dark, sunburnt people, scorched with the sun; full of wit and animation, always in love, always making verses, listening to music, having galas, dances, and tournaments." A good start to a dramatic space opera perhaps, but the merchandising rights were presumably not very lucrative.

Now, before we get to Hollywood-worthy aliens accompanied by awesome action figures, it's important to take stock and note that, before the Industrial Revolution, extraterrestrial beings were only imagined as superficial variations of Earthly life, rather than the truly alien entities we seem to prefer today. Remember that, ask Siri to take a note, or just bookmark this page because it's going to be important later on as it helps highlight the stupidity of alien fantasies.

Mars attacks

As the lanterns of the Industrial Revolution lit up the nineteenth century, our view of the cosmos was still fuzzy, but a dramatic shift was happening down on Earth. Progress in science and technology led to a cultural shift in the belief that the universe was not just the playground of gods but a vast expanse awaiting human exploration. And as our real-world boundaries expanded, so did our fictional ones.

We set our telescopes on celestial bodies to reveal detailed patterns and structures. Observations of Mars were particularly captivating. In 1877, Italian astronomer Giovanni Schiaparelli reported seeing *canali* on Mars. While *canali* means something like *channels* in English, without constant reminders from the Duolingo bird, it was misinterpreted in the English-speaking world as *canals*. The assumed existence of canals on Mars led to widespread speculation that these were artificially constructed waterways on the planet created by intelligent Martians.

The idea of an ancient and possibly dying Martian civilization desperately building canals to harness water from the polar ice caps captured the public's imagination. This theory was popularized by American astronomer Percival Lowell, who wrote several books on the subject. Lowell's observatory in Arizona became a hub for the study of Mars, and his observations and speculations fueled the fire of Martians as advanced, intelligent beings.

Technological advancements on Earth, such as trains, telegraphs, and, well, canals, were transforming societies at unprecedented rates. If a Martian civilization could do this on a global scale, they would be far more advanced and far more coordinated. And, hey, look at that nice blue planet over there with lots of water...

Enter H. G. Wells with his groundbreaking novel, *The War of the Worlds*. This story, published in 1898, imagines a Martian civilization that, instead of building canals to save its own planet, sets its sights on ours. But Wells's Martians were not humanoid. They were octopus-like creatures with vast intellects who used advanced technology, including giant fighting machines, chemical weapons, and heat-ray guns. While he didn't call them tanks, mustard gas, and lasers since they didn't exist yet, Wells was the world's first futurist, essentially "inventing" technology and predicting its use in war decades before it came to pass. He also foretold conflict and disease on a global scale. Spoiler alert: The Martians basically succumbed to a pandemic. If this guy also didn't invent time machines, I'd be suspicious!

The Martians of Wells, imagined by Steven Spielberg in 2005 as the result of E.T. calling his drug dealer instead of phoning home, contrasted with the romantic notions of peaceful humanoid extraterrestrials with no interest in Earth. It was the first work of fiction to explore the consequences of an encounter between mankind and beings visiting from another world. It set the stage for future depictions of aliens not just as curiosities but as existential threats. This theme has been revisited countless times in literature, films, and television shows, establishing a new, more antagonistic paradigm for human-alien interactions.

It's incredible to think that all of this, from epic sci-fi to tinfoil hats, was due to a mistranslation. I like to imagine what the world would have been like if *canali* had been mistranslated as *cannoli* instead. Maybe instead of fearing extraterrestrial invasions, we'd be tuning in to Gordon Ramsay blasting off to Mars in search of soggy bottoms. The suspense of a seven-month journey, all to determine whether Mars is, in fact, cake, is something the universe will sadly never experience.

A Golden Age of Aliens

Apparently, anything can have a "golden age." In literature, there was a time when "the classics" were just "the books." The world of art had its various movements, and film had a time when not every movie was a reboot, sequel, or based on a comic book. There were even golden ages in the world of piracy, when the Jolly Roger flew proudly over rum-tainted Caribbean waters. Now, if anything, it seems we're living in the golden age of conspiracy theories, where you're never more than a click away from going down a rabbit hole filled with lizard people microchipped with mind-control vaccines personally made in a pizza shop by Bill Gates.

Joking aside, since golden ages are meant to be the *peak* of something destined for a sharp decline, I really do hope we are currently living in the Golden Age of Influencers. Good riddance. Anyway, in the official history of American science fiction endorsed by the gatekeepers of Wikipedia, the Golden Age of Sci-Fi comprised the years 1938–1946.

This was during a time when an explosion of magazines printed on cheap pulp paper—so-called pulp fiction—were crammed with terrible detective stories, lurid erotic tales, and extraterrestrial fantasy. While generally considered venues for low-quality literature, these magazines laid the groundwork for the sci-fi genre's modern tropes. Indeed, one of the most famous sci-fi authors of all time, Isaac Asimov, began his career writing for such magazines. Variously referred to as the greatest science-fiction short story, "Nightfall" appeared in a 1941 issue of *Astounding Science Fiction*. The story portrays a planet with six suns that constantly bathe it in sunshine—except, of course, once every few thousand years when an eclipse casts darkness on half the planet and society collapses.

The Golden Age of Sci-Fi was essentially the Wild West of extraterrestrial storytelling. We embarked on intricate tales of space operas where empires were more vast than your grandmother's collection of porcelain cats. We met some aliens just wanting to borrow a cup of sugar and others hell-bent on redecorating Earth with plasma fire. There were even artificially intelligent robots, questioning their existence, recounting utopian dreams, and generally just stoking the angsty blend of hope and fear toward a tech-driven future. All the alien narratives were seeded here. But with the events of World War II, humanity stumbled into a new age where, honestly, sci-fi writers probably felt a bit upstaged by reality.

Are we the bad guys?

Having mastered the art of annihilating ourselves, the obvious next step was to take the shit show on the road. The world gawked at the destructive power of rockets and bombs and logically concluded that instead of decommissioning them, we should just aim them up instead of down. Hence, the space race was born, turning our technology into an odd mix of starry-eyed dreams and political dick-swinging, which was never as impressive as it sounds. In the science-themed corners of fiction, things were getting equally dicey, if perhaps a little less consequential.

Robert A. Heinlein's 1950 novella *The Man Who Sold the Moon* tells of an eccentric businessman driven by an insatiable dream to make the first manned Moon landing, not out of scientific curiosity or geopolitical concerns but as a commercial venture. While we know that's not how the actual Moon landing in 1969 went down, there is still time for Mars. If only we had a rich, megalomaniac tech CEO deeply influenced by a belief in the boundless potential of private enterprise...

The tales of extraterrestrial encounters and spacefaring adventures weren't confined merely to our solar system or even the written page. Sci-fi narratives leaped from pulp magazines and novels onto the silver screen and television sets that were growing in both popularity and size. "Seeing is believing" was a dangerous precedent as depictions of alien visitors reached the eyeballs of an ever-growing audience...with ever-shrinking brainpower. The mixture of serialized television, space-age dreams, Cold War anxieties, and the inability to recognize a man in a rubber suit resonated especially deeply with U.S. audiences, embedding the genre's tropes into the very fabric of popular culture.

The aliens that stumbled onto our screens in those days looked more like the staff at a thrift-store Halloween party. In *The Day the Earth Stood Still* (1951), our extraterrestrial visitor stepped off its cardboard flying saucer, looking like one of the honorable mentions at a RoboCop cosplay contest. Official contender for "the worst film ever made," the 1953 *Robot Monster* starred a man in a gorilla suit with a diving helmet on. Even "acclaimed" pictures such as the 1953 cult classical *It Came from Outer Space* featured a spaghetti monster with an eyeball for a head conveniently sized to human proportions.

The postwar era also witnessed the launch of human space exploration on galactic scales. Starting with the original series in the 1960s, the ongoing Star Trek universe focuses on the adventures of the USS *Enterprise* as its crew explore new worlds and meet diverse alien species, all in the service of the copyright theft of memorable catchphrases like "live long and prosper," "resistance is futile," and "That's not where that goes, Captain Kirk."

Cosmic Cambrian explosion

In the last several decades, extraterrestrial science fiction has spawned a diversity of species only limited by the number of people's names that can fit within twenty minutes of credit roll. Computer-generated graphics (CGI) heralded a new era where storytellers were no longer constrained by practical limitations of whatever could be found in the prop room. Instead, aliens could look as wild as an acid trip or as realistic as your next-door neighbor—you know the one I'm talking about.

Modern blockbusters like James Cameron's *Avatar* brought to life the entire planetary ecosystem of Pandora, where creatures and plants connect to a global biological internet with a latency that could only exist in the wet dreams of Comcast executives. The sequel, *Avatar: The Way of Water*, with a budget rivaling the gross domestic product of a small nation, is one of the most expensive films of all time, which is the price to pay for hyper-realism and ego maintenance. Ironically, though, the more "realistic" these aliens become, the harder it becomes to suspend disbelief. In its pursuit of high-definition precision, CGI inadvertently deprives us of the gaps our imaginations once eagerly filled.

Books and pre-CGI films, with their grainy, low-sophistication charm, offered a canvas only partially painted. The slower pacing, longer shots, and ambiguous endings forced us to co-create the narrative. The silhouettes in the mist, the eerie noises just off-screen, the suggestive dialogues—they gently nudged us, forcing our minds to wander, wonder, and weave tales far more imaginative than those appearing on the screen. In those dimly lit theaters, each viewer, armed with their personal set of fears, hopes, and experiences, sculpted their unique extraterrestrial first-encounter scenario.

The realm of science fiction has always operated as a mirror to humanity's dreams, anxieties, fears, and aspirations. But as the genre's popularity skyrocketed, it started to look more like the Evil Queen's magic mirror than a passive reflection of our psyche. What was once relegated to fantasy in the tales of Wells and others became whispered as reality. Indeed, the magic mirror compelled some not only to *believe in* but to *see* aliens—or at least believe they saw aliens. Yet, this was *not* the product of a Michael Bay chain reaction of hyper-realistic explosions. Oddly enough, it was from the plastic saucers held up by strings.

The Grays

So far, we have been discussing intentionally fictitious depictions of aliens. As you are most likely aware, there are now many *un*intentionally fictitious stories of extraterrestrial encounters—and, depending on your particular fetish, you've waited far too long for the kinky stuff.

Quick! Picture an alien in your mind. Unless you're my racist uncle after too many rum and eggnogs during the family Christmas party, I bet I know what you are imagining. Interestingly, the most popular image of an alien—oversized head, noseless nostrils, large almond-shaped eyes, and a slender gray body—does not come from fiction at all. These so-called "gray aliens" have become one of the most widely recognized representations of extraterrestrial beings in popular culture.

The iconic image of the Grays is largely due to reported alien abduction cases in the midtwentieth century. The most memorable early instance was the Betty and Barney Hill "abduction" in 1961. The Hills, a married couple from New Hampshire, claimed they were taken aboard a flying disk vehicle by extraterrestrials and subjected to the kind of examinations which became the defining example of the term *anal probe*. Seeking to stretch their fifteen minutes of fame to hours of primetime specials, the couple basked in media attention. Their interviews, photographs, and hypnotic regression sessions were extensively covered, providing the perfect template for extraterrestrial encounter narratives. The accounts of their examination aboard the alien craft involved extensive probing of their bodies, with a particular emphasis on the fun bits.

The bulbous heads of the Grays were further etched into

public consciousness through art imitating life with hits like Steven Spielberg's 1977 blockbuster *Close Encounters of the Third Kind*. Meanwhile, the other end (no pun intended) was solidified as a permanent aspect of extraterrestrial mythos courtesy of a blend between the rise of niche internet forums and cable television's relentless pursuit to see who could cause the most offense to prudish conservatives. Case in point: *South Park*'s 1997 series premiere, titled "Cartman Gets an Anal Probe."

While nearly every Gray, purported or fictionally depicted, arrives on Earth in some vehicle, the opposite is not necessarily true. Think it of like afternoon traffic. Everyone in the parking lot gets into a car, but can we assume every car that ends up on the road has a person in it? Presumably, yes. By the same logic, we seem to assume that every floating or flying thing we see or imagine in the sky is piloted by something that looks like Steve Buscemi playing a rotten celery stalk. And, holy shit, do humans ever see a lot of spooky stuff in the sky.

Flying dinnerware

In ancient times, reporting mysterious lights or shapes in the sky to the authorities meant a trip to the temple, where, of course, you'd get back arbitrary divine omens and a prescription to sacrifice some children. Fast-forward a few centuries, and, at least in Europe, you wouldn't want to mention the "wisps" you saw over the marsh lest you wanted to waste the rest of your evening as part of a torch-and-pitchfork rally. As we got a grip on industrialization and made flying machines of our own, sightings started having a more human-constructed appearance. The "mystery airship" craze in the late 1800s saw numerous reports about mysterious blimps, which now sound more like plot devices of an awesome steampunk story rather than the actual newspaper speculation of a Martian invasion.

Things started to take off during World War II when both Axis and Allied fighter pilots reported seeing fiery orbs that trailed or flanked their planes. Dubbed *fuckin' foo fighters* and later sanitized for media consumption as *Dave Grohl*, reports were initially suppressed to avoid panic and misinformation. It's no secret that wartime matters are kept secret, which didn't bode well for a country that *loves* a good government conspiracy theory. The military's concern was obviously that foo fighters were secret enemy weapons, but eventually they were chalked up to atmospheric oddities since pilots from all countries reported them, and none apparently posed a threat. So far as we know, no one's been killed by a foo fighter, and since Dave Grohl claims to be the only musician not regularly high on cocaine and heroin, I suspect we are pretty safe.

There was no immediate association of foo fighters with extraterrestrials, but it just fits the whole government cover-up narrative so well that you would be forgiven if you believed it did after a brief internet search today suggested otherwise. The timing was too perfect as, postwar, with rocket science and space exploration capturing the public's attention, the stage was set for the flying saucer frenzy. A particular sighting in 1947 by Kenneth Arnold is often cited as the origin of the term *flying saucer*, which upstaged *foo fighter* as the default term for any unexplained sighting.

The Arnold incident—or, more accurately, the media coverage of Arnold's account—launched the "flying disk craze" of 1947 when thousands of similar reports were made, including the far more famous "Roswell incident." After an unidentified object crashed on a ranch near Roswell, New Mexico, initial reports called it a *flying disk*. Military officials later clarified, with photographs and all, that it was a weather balloon. Despite the official account, speculation and theories abounded, and Roswell became synonymous with aliens and government cover-ups.

Unfortunately, Roswell today stands as a stark testament to humanity's unerring capacity to commercialize, trivialize, and turn any semblance of mystery into a blatant cash grab. Once echoing the vibrant hum of those idyllic 1950s towns you see in period dramas, its streets now exude an ambiance akin to wandering through the abandoned carnival set of a Stephen King movie. Storefronts peddle "authentic" spaceship parts alongside T-shirts that read, "I was probed in Roswell." An unsuspecting visitor might be forgiven for mistaking this crash-landing site for the abduction site of humanity's collective dignity.

UFOs

The term *unidentified flying object* and its initialism were coined in the 1950s by the U.S. Air Force to refer to any airborne phenomena that couldn't be easily identified. It replaced the more sensational term *flying saucer*, and its bureaucratic blandness was no accident. It was meant to distance these sightings from the growing public hysteria surrounding extraterrestrial life. However, the public had other ideas, and as the Space Age dawned, UFO sightings only became more frequent and increasingly intertwined with suspicions of alien encounters.

The second half of the twentieth century was rife with tales of strange objects in the sky and an ever-growing list of abduction stories. The modern era of self-styled "ufologists" was ushered in with gusto. The 1960s through the 1990s saw a surge in reports, supported by a deluge of grainy photographs and shaky camcorder footage. The stories varied from glowing orbs, metallic saucers, triangular craft, and the ever-elusive giant floating penis euphemized as a "cigar-shaped object."

These decades were the heyday for a burgeoning UFO subculture complete with conventions, books, documentaries, and, of course, numerous conspiracy theories. Whispers of shadowy government organizations conducting cover-ups became as widespread as the sightings themselves. Near-religious movements formed around the belief that governments, especially the U.S. government, not only had proof of extraterrestrial visits but also were keeping live aliens at sites such as Area 51, an actual military base in Nevada.

But as we transitioned into the twenty-first century, something curious happened. With practically everyone carrying a high-definition camera in their pocket, thanks to smartphones, you'd think UFO evidence would be overwhelming. Yet, while images and videos of purported UFOs certainly proliferated, so too did the means to debunk them. Photo manipulation software, drones, and an increasingly skeptical public armed with instant access to information meant that genuine-looking UFO captures were swiftly explained or exposed as hoaxes. Instead of a boom in clear evidence, we got a lot of wasted pixels and yet even more hand-drawn pictures of shapes and lights in the sky.

The unfortunately official-sounding National UFO Reporting Center, a nongovernment business, allows anyone to submit reports of UFO sightings on its website. Besides the sometimes-hilarious testimony accompanied by kindergarten-level artwork, the only interesting thing about these reports is that they seem to peak at the same time any other conspiracy theories do, such as during every presidential campaign season, during the peak of the COVID-19 pandemic, and during recent UFO-related congressional hearings.

Want to believe

With the rise of the internet, UFOs have gone from late-night bar conversations to being fodder for meme warfare and viral challenges. "Storm Area 51, They Can't Stop All of Us," for example, wasn't so much an earnest battle cry as it was a testament to the power of modern folklore intersecting with online absurdity. Millions of clicks later, what started as a Facebook joke event ended up being a real-world gathering. No aliens were liberated, but plenty of merch was sold, proving yet again that the truth might be out there, but it'll probably be on a T-shirt.

One in four Americans claims they've seen a UFO, which could be anything from floating lights in the sky to zipping saucer-shaped shadows. One-third of this group believe extraterrestrials are responsible for these purported UFOs. But take these numbers with a grain of salt—or a swig of moonshine—because the same poll suggests that a significant chunk also believes in Bigfoot. A separate study found that half of Americans suspect that the government is hiding information from them about UFOs, which is probably true but also has nothing to do with aliens.

But don't think that I'm picking on Americans here. It's just where most of the media attention is. In fact, global polling data echoes these numbers. A seemingly serious survey from late

2022 trying to forecast short-term future trends asked a bunch of boring questions about healthcare, economics, and whatnot but also included a question about whether the participants thought that Earth would be visited by aliens in 2023. Remarkably, only two-thirds of participants thought this was unlikely. And if that doesn't boggle the mind, consider that in 2010, a Reuters News poll found that one in five respondents believed aliens were already among us, masquerading as humans. Perhaps they're the ones taking these polls, which would explain a lot.

While the stories today are far more varied and receive little to no traditional media attention, belief in aliens does not seem to have waned over the years. There are also theories that stretch incredulity to its breaking point. Shows like *Ancient Aliens* and similar mockumentaries not obviously presented as parodies only confuse audiences into thinking self-styled rogue scientists have even a hint of credibility. We could go down the rabbit hole of moronic ideas like interdimensional beings, aliens living on the other side of a flat Earth, lizard people, or whatever is in Tom Cruise's browser history, but, frankly, dedicating more than a paragraph to such stupidity is a disservice to our remaining brain cells. Let's get *serious*.

Taking things seriously

The U.S. government has stated on numerous occasions that it takes UFO reports "seriously," unlike most rational people who simply ignore them and those who go a bit further to ridicule them. The first official investigation by the Air Force was Project BLUE BOOK, which collected 12,618 UFO sightings and wrapped up in late 1969. The summary was unequivocal—there was no evidence of a threat to national security, no evidence of advanced terrestrial technology, and no evidence of extraterrestrial origins. In addition to calling bullshit, the report concluded that sightings were either conventional objects, deliberate hoaxes, or the ramblings of insane people.

As the decades rolled on, the U.S. government's official position was a disinterest in UFOs. However, in 2007, the obviously classified Advanced Aerospace Threat Identification Program (AATIP) was born. While its $22 million budget from 2007 to 2012 is probably less than the rounding errors in the total Defense budget over the same period, its mere existence proved irresistible for conspiracy enthusiasts when it was revealed in late 2017 with predictably little detail. "A secret government program dedicated to UFOs? They must know something!" became the rallying cry.

With its love of all things pointlessly bureaucratic, the U.S. Department of Defense (DoD) then rebranded and revived the project in 2017, donning the new title Unidentified Aerial Phenomena Task Force (UAPTF). And if AATIP and UAPTF sounded too mundane, the Defense Intelligence Agency (DIA), ever the fans of the TLA (three letter acronym), introduced the Airborne Object Identification and Management Synchronization Group (AOIMSG) in 2021, eventually transitioning to the All-Domain Anomaly Resolution Office (AARO) in 2022. If their goal was transparency and less confusion, they've clearly failed. My own conspiracy theory is that the U.S. government has an extraterrestrial troll in some basement level of the Pentagon who comes up with these acronyms.

In any case, Sean Kirkpatrick, the director of AARO until late 2023, provided a statement on the preliminary findings, which could have been predicted by a single neuron. He said, "The majority of unidentified objects reported to AARO demonstrate mundane characteristics of balloons, unmanned aerial systems, clutter, natural phenomena, or other readily explainable sources." In other words, all these programs, acronyms, and taxpayer dollars have been spent on reassuring people that their UFO fantasies consist mostly of birds, trash, or the result of staring too long at the Sun.

To be absolutely clear, a military agency tasked with defending the borders of its nation *has to* take sightings seriously. Could you imagine otherwise? When your risk tolerance is essentially zero, it doesn't matter if it is a shape on the horizon or something floating in the ocean or sky—all reports must be assessed. The government's very seriousness about the issue is seen by many as undeniable evidence that UFOs, or UAPs, or whatever the basement-troll branding consultants are calling them these days, must be aliens. But for all the hundreds of thousands of reports, the official response from anyone who takes this more seriously than "I can't immediately explain it, so...aliens" is this: *There is no evidence of extraterrestrial origin.*

For the rest of this chapter, then, we are going to assume *not* that aliens don't exist but that we currently have zero evidence that they do. That being said... [Editor's note: Several paragraphs of tantalizing evidence and testimony have been redacted and appear in the classified version of this book.]

Same same, but different

So, without evidence of aliens, how does one explain all the sightings? Picture ancient Sumerians whispering tales of gods descending from the heavens in golden chariots. Or the Greek myths of deities soaring across the sky, occasionally popping down to Earth for a booty call. These narratives are eerily reminiscent of modern tales where extraterrestrial beings descend upon our world in gleaming ships for similarly lewd activities. (Though no one told modern aliens that's not how babies are made.) The arc of extraterrestrial stories never changes; the only difference is the supposed technology being used. First, it was chariots and boats, then kites and blimps, and finally airplanes and spaceships.

Why does the Greek god of blacksmithing need to fashion Achilles some armor instead of Zeus just air-dropping it to him from a spaceship? The answer is simple. It doesn't matter the means through which Achilles enacts his revenge, but it must match the cultural milieu of the time. If the *Iliad* had first been written today, it wouldn't involve swords, shields, and chariots but guns, spaceships, and Chris Hemsworth's abs. Indeed, "Achilles" is basically the protagonist of every action film. For example, Marvel's reimagining of Thor has him gifted a powerful weapon of extraterrestrial origin, a symbol of his worthiness, which he must prove in a heroic journey of humility, self-sacrifice, and epic poses while a camera pans around him at a low angle.

UFO sightings and alien encounters are but modern manifestations of the age-old storytelling tradition. We are, by nature, storytelling animals. Ever since our ancient ancestors discovered and touched a large black monolith that magically gave us consciousness, we've used tales and myths to understand the world. Every society has its own set of stories used to structure its culture, maintain cohesion, and guide the actions of its members. In the vast expanse between conspiracy theories and cutting-edge science lies a complex web of human psychology and culture that acts as the environment in the Darwinian natural selection of cultural memes. Those that survive do not do so because they are *true* but because they are easy to share.

In this vein, modern technology has also brought with it the behemoths of consumerism and sensationalism. Alleged UFO sightings no longer die in obscurity shortly after they are spoken. Instead, they are commodified, plastered on merchandise, turned into documentaries, sold as breaking news, and rocketed to the top of social media feeds. Roswell isn't just a town in New Mexico; it's a brand, a tourist attraction, a T-shirt design. In short, it's a shithole—a shithole that's found a niche in the ever-more-fluid global ecosystem of publicly aired thoughts.

Ideas that were once at the fringes of rational discourse now find their way into mainstream discussions with alarming ease. Like sprinkling some measles at an anti-vax rally, an unsourced tweet now spreads faster than it takes to sit down in the effort to write a researched article. We've engineered an environment where bullshit thrives, and the only solace we have is that, like the trillions of bacteria we carry around, we at least remain useful vessels for it.

Out-of-this-world cure

The antidote is, of course, well known and comes in two flavors: sweet and spicy. Carl Sagan, astrophysicist and science communicator par excellence, was known for his fascination with the idea of extraterrestrial life. Yet, even he famously said of it, "Extraordinary claims require extraordinary evidence." This assertion became the beacon of rational thought and skepticism. The more outlandish a claim, the stronger the evidence needs to be to support it. It's a principle that skeptics brandish, ironically not unlike a mythical sword, against the myriad claims of UFOs, alien abductions, and the ever-growing pile of bullshit seeping out of the internet.

If Carl Sagan was the Mr. Rogers of science communication, Christopher Hitchens was the Simon Cowell. The sharp-tongued journalist and prototypical contrarian echoed Sagan with a similar sentiment now known as Hitchens's razor: "What can be asserted without evidence can also be dismissed without evidence." In a world rife with tales of abductions, sightings, and probing, Hitchens's razor is the spam filter. Outlandish anecdotes are as much evidence of aliens as an illegible email offering you $4.3 billion is evidence that a Nigerian prince has your contact details. I mean, what would it really take for you to believe an alien crash-landed in Roswell? Surely, it's more than government photos of a torn-up balloon, folklore, and the incoherent nonsense in the brochure of a town reliant on incredulous tourists.

For the sake of transparency, let me share a little secret. Just a moment ago, I experienced a UFO encounter. I stepped outside, stared at the expanse above, closed my eyes, and pressed firmly on my eyelids. Lo and behold, dancing lights materialized. The logical thing should have been a search for an explanation online, but jumping to "it's aliens" sounds more fun. Alternatively, I could just chalk it up as just another insignificant blip in the unbroken chain of unexplained clicks, ticks, lights, and smells that we typically ignore subconsciously. My point here isn't to belittle genuine experiences but rather to emphasize that while seeing is believing, our believing in something imposes nothing on reality. That it is so easy to believe ought to be cautionary. It calls for discernment, grounding our beliefs in tangible evidence rather than sensations our mind conjures into illusions.

Aliens probably, maybe exist

There's actually a science of aliens called *astrobiology* and we'll get to that soon. But given everything I've just said, you might wonder why such a thing exists at all.

Enrico Fermi is said to have posed the infamous question, "Where is everybody?" during a casual lunch conversation with other physicists at the Los Alamos National Laboratory in the summer of 1950. It's a beautiful question because every answer is equally interesting...or terrifying. Either we find out where other life exists, or we figure out why we are the only ones here.

The universe is nearly fourteen billion years old, and our Milky Way galaxy has been around for almost as long. With such a huge time frame, even a slow-moving civilization could have traveled the galaxy multiple times. But we've found nothing—no physical encounters, no signals, not even any bags of shit left behind. Though, this assumes an intelligent civilization would leave bags of shit on every celestial body they visited. I'll pause and let you google it. These contradictory alternatives define what has become known as the *Fermi paradox*.

The *Drake equation*, proposed by astronomer Frank Drake in 1961, quantifies the Fermi paradox. It's an attempt to break down the complex question of the existence of alien civilizations into a series of smaller questions, each representing more specific quantities that are easier to estimate. The equation takes into account factors like the rate of star formation, the fraction of stars with planets, the fraction of planets that could support life, the fraction on which life actually appears, the fraction that evolves to intelligent life, the fraction that releases detectable signals, and so on. Many of these figures are unknown and can only be pulled from one's ass, which is what Drake himself did with his initial estimate of ten thousand other advanced civilizations in our galaxy.

Oh, what's that, you want to do some math? Okay, if you say so. Let's start with the rate of star formation in our galaxy. This tells us how many new stars are born each year, potentially harboring planets for life. Forming a brand-new nuclear furnace with the average size of our Sun is not a trivial feat. Our galaxy births about two new stars per Earth year. Next, we need to know the fraction of stars with planets. Not all stars have planets, just like not all planets have moons—poor Mercury! Exoplanet searches often come up with planetary systems, so let's toss a coin and say 50 percent of stars have planets.

Okay, so we are at one new planetary system in the galaxy per year. But not all planets have life, as evidenced by our own cosmic neighborhood. So, we need to know the number of planets that could support life. We're still holding out for Mars, so let's be optimistic and say there are two planets that are the right size and temperature to hold liquid water. But just being in the right place doesn't guarantee life can develop even if it gets started. We need to know the fraction of potentially habitable planets where life actually develops. This is obviously the big unknown since we don't even know how life got started here. Again, let's be optimistic and say 10 percent.

We are now imagining planets where life gets going. Even if

life arises, it's unclear how often it evolves intelligence like ours. The evolution leaps that accumulated over billions of years are staggering. Let's go with a modest estimate that 1 percent of planets that harbor life will evolve intelligence. But we had intelligent life on Earth for millions of years—and a great deal more if you include non-humans—and only recently started airing our grievances and trivia into the cosmos. Pulling the old number-out-of-ass trick, let's guess 10 percent of intelligent civilizations develop a technology that releases detectable signals.

Last but not least, we need the average lifetime of a technologically advanced civilization. This depends not only on natural disasters, like what happened to those genius dinosaurs, but also on factors like self-destruction. Clearly, now is the time for pessim...I mean, realism. Suppose, and may I remind you that we've yet to prove this wrong, we give them a full millennium. Multiplying all these numbers together, the Drake equation estimates ten thousand technologically advanced civilizations with teenagers recording and transmitting their farts over the airwaves in our galaxy at any given time.

In other words, the Drake equation gives us a hopeful picture of a shopping mall galaxy buzzing with conversation. On the other hand, the Fermi paradox poo-poos all that, introducing a dose of empirical skepticism suggesting the galaxy is more like an abandoned parking lot. Both the equation and the paradox stir up much discussion, but in science, we focus on what can be measured, tested, and quantified. Following Drake's lead, scientists break the big question down into smaller ones and then force a bunch of graduate students to answer them in a totally fair trade of the best years of their lives for a two-hundred-page thesis no one will read. And, while that sounds drab, a quantified, overly pedantic answer to a small, obscure question is still far more valuable than qualitative bullshit fit only for tabloids.

Alien science

Its name coined in the middle of the twentieth century, *astrobiology* is the science of aliens. While wisely not using the term *aliens* in any of its textbooks, this discipline generically refers to the systematic study of life in the universe, including its potential origins, evolution, and distribution. Much of astrobiology is characterized by actively searching the cosmos for life rather than waiting around for people who can't work a tripod to report on erotic fever dreams. In our search, we look for *biosignatures*, the technical term for "signs of life."

With biosignatures, the basic idea is not to find life itself but to find the smoking gun that indicates life. We hope to detect by-products of existing life or footprints left by past life, which might not be "alive" themselves. As you might imagine, this becomes nuanced very quickly. For example, life on Earth produces a lot of simple molecules like oxygen, carbon dioxide, water, methane, and so on. But these are also common on other planets right here in our solar system in places we're pretty sure contain no life. Venus, for example, has an atmosphere almost entirely of carbon dioxide. But none of the dozen or so probes we've sent there have lasted more than two hours in the crushing heat and pressure of its atmosphere, painting a slightly different picture than Fontenelle's vision of a sun-soaked perma-festival. For now, then, we have settled on searching for highly complex molecules that are unlikely to have arisen from directionless nonbiological processes or simply by chance.

The direct search for biosignatures is supported by many secondary factors. For example, life on Earth requires oceans of water, so there isn't much sense in looking for living things where there is no water. We stopped sending things to Venus but still hope to find evidence of past life on Mars, which currently has ice under its surface but is widely believed to have held liquid water in the past. The beds of ancient Martian lakes and streams are presumably a good place to look for biosignatures. Indeed, there are currently three active rovers on the planet. As I write this in 2024, the *Perseverance* rover is caching Martian rock samples for a future return mission a decade from now. I would have called the rover Sisyphus, but no one asked me.

In the following decades, proposed missions will have us looking up close at some of the moons of Jupiter and Saturn, first to search for liquid water that is currently believed to lie beneath their surfaces. Unfortunately for me, that's probably it. I'll be long dead before humanity or its robotic progeny venture to other solar systems. On the bright side, we do have telescopes that can take awesome photos of our galaxy and beyond. While we've been able to convincingly detect planets outside our solar system—called *exoplanets*—since the 1990s, it wasn't until the Kepler space telescope launched in 2009 that we could say beyond reasonable doubt that other "Earth-like" planets existed.

According to the NASA Exoplanet Archive, as of October 24, 2023, we have confirmed exactly 5,535 exoplanets, with Kepler finding 361 planets within the *habitable zone* of its host star, meaning the temperature would be just right for liquid water to exist on the surface. Since the Kepler mission could only cover a tiny fraction of the sky and a small number of stars, extrapolation leads to an official guess of about eleven billion exoplanets in the habitable zones of Sun-like stars in our galaxy alone. Surely, one of them couldn't be much worse than Missouri.

Kepler's successor, the Transiting Exoplanet Survey Satellite, TESS, as it's better known to a few dozen people still watching, was designed to spot planets around bright, nearby stars in the hopes of finding targets for direct imaging by the most recent object of space-nerd obsession, the James Webb space telescope (JWST). The JWST is the biggest, baddest, gold-plated, honeycomb-mirrored space technology we've ever made. As of this writing, the published results are mostly weird flexes for press releases, but its capability of direct imaging of exoplanets and the analysis of their atmospheres has been demonstrated. This is a book and not a living document, so I'm going to have to stick my neck out and make a wild guess. You probably won't be surprised that I'll play the role of curmudgeonly prophet and say that, despite its awesomeness, the JWST will not find convincing evidence of extraterrestrial life.

What is life anyway?

It's not all fancy press releases and glossy magazine covers full of optimism in our quest for alien brethren. Our hunt for extraterrestrial life is somewhat hindered, not entirely by a lack of imagination but by a surplus of academic caution. We're essentially the drunk searching for our keys under the lamppost because that's where the light is. If we stumble too far, we might end up tripping into a pile of dog shit or, worse, end up being interviewed by the History Channel, which would be the ultimate embarrassment. Thus, our search is largely based on "life as we know it." However, considering the vast and bizarre array of conditions on distant worlds, this might be shortsighted.

Much like the filters on an Airbnb search, we look for water, a comfortable range of temperatures, and a nice healthy spread of organic molecules. Yet, more akin to the results of an unfiltered Tinder search, life could be lounging in the methane lakes of Titan, the sulfuric acid clouds of Venus, or New York City studio apartments. We operate on very human-centric notions, which are hard to break away from. Some research here on Earth attempts to move in the other direction. So-called *extremophiles* are organisms that thrive in conditions considered inhospitable to most life forms. For instance, thermophiles flourish in the extreme heat of boiling geyser water, halophiles thrive in high-salinity environments like salt flats, and toxicophiles manage to survive in the comments section of internet forums, mostly by catalyzing self-righteousness through their keyboards. All these extremophiles stretch our understanding of what a "habitable zone" even means, extending the possibilities for life to a broader range of celestial objects.

We look in the light because if we had to answer questions like "What is life?" first, we'd never get started. A question of similar difficulty is "What is intelligence?" Luckily, any definition will include the ability and need to exploit the laws of physics, a trait we share with a hypothetical advanced civilization, no matter how foreign it might be. Signs of technology use are called *technosignatures*. Earth, for example, ought to be giving off such signals, at least for the past few centuries, which is enough time to have reached thousands of nearby stars.

Sending out an SOS

Signals from Earth bear the hallmark of our technological

advancement, painting a picture of a civilization that has grown adept at harnessing the electromagnetic spectrum, if only for the purpose of meme sharing. Among the cacophony of signals we've unleashed into the cosmos, our use of radio waves for wireless communication is a source of ever-growing cosmic chatter. From the early radio broadcasts to the ceaseless hum of modern data transmissions, we've essentially turned our planet into a megaphone of banal status updates interspersed with advertising. If there's anyone (or anything) tuning in, our penchant for blasting our every thought into the aether might be the first hint of our existence.

The illumination of the dark side of our planet is another *glowing* sign of our technological stride. The sight of Las Vegas from space might attract alien civilizations like moths to a free buffet or video poker machine. The atmospheric by-products of our industrial processes, though not visible to our own eyes, unless you live in Long Beach, are also a significant indicator of our technological activities. And let's not forget the *pew-pew* of the lasers we employ in scientific research, military applications, or just shoot into space for fun. Their coherent beams are capable of traversing the void with minimal losses.

Among all this cosmic graffiti, we have also sent several intentional messages into the void. In other words, the fraction of known intelligent civilizations in the universe that send out radio greetings is one, which is an excellent statistic to throw out when you want to convince potential funding agencies of your technosignature research.

The *Arecibo message* was our first attempt—sort of—at interstellar messaging. Crafted by a team led by none other than Frank

Drake and Carl Sagan, this message was broadcast in 1974 from the Arecibo Observatory in Puerto Rico—the very same facility that Pierce Brosnan—or possibly his stunt double—famously danced across to save the day in *GoldenEye* to prevent a similar message being sent. The message comprises only 1,679 binary digits (which other nerds will also recognize as a prime number). I quote:

"0000001010101000000000000101000001010000000
1001000100010001001011001010101010101001001
0000000000000000000000000000000000001100
0000000000000000110100000000000000000000011
0100000000000000000010101000000000000000000
0111110000000000000000000000000000000011000
0111000110000110001000000000000001100100001101
0001100011000011010111110111110111110111110000000
0000000000000000000100000000000000000001000
0000000000000000000000001000000000000000
0011111100000000000000111110000000000000000000
0000011000011000011100011000100000001000000000
00100001101000011000111001101011110111110111110111
1110000000000000000000000000000001000000110000
0000010000000000011000000000000000010000011
00000000000111111000001100000011111000000000001
1000000000000010000000010000000001000001000
0001100000001000000001100011000000100000000
0011000100001100000000000000011001100000000
00000110001000011000000000011000110000001000
0000100000010000000010000010000000011000000

00100010000000011000000001000100000000001000
00001000001000000010000000100000001000000
00000110000000001100000001100000000100011
10101100000000001000000010000000000000100
00011111000000000001000010111010010110100000
01001110010011111101110000111000001101110000000
00101000001110110010000001010000011111100100000
00101000001100000010000011011000000000000000
00000000000000000000011100000100000000000
00111010100010101010101001110000000001010100
00000000000000101000000000000000111110000000
00000000011111111100000000000011100000001100
000000110000000000011000000011010000000001
01100000110011000000011001100001000101000010
10001000010001001000100100010000000010001010
00100000000000010000100001000000000001000
00000010000000000000010010100000000000011110
01111101001111000."

Even after being proofread several times, that's still more interesting than an Elon Musk tweet.

The Arecibo message was a binary encoding (the ones and zeros of computer-speak) of some random facts like the numbers 1–10, the atomic numbers of the atoms in DNA, the size of the human genome, and so on. Clearly, it *means* something to us— provided someone told you how to read it—but it was hoped that a sufficiently intelligent recipient could also decipher it. In any case, the message was aimed at Messier 13, a cluster of stars twenty-five thousand light-years away, making it more of a symbolic gesture than a practical attempt at communication. Indeed, the Arecibo Telescope collapsed in late 2020 and although Google Maps claims it is merely "temporarily closed," there are no current plans to rebuild it—not for at least fifty thousand years anyway.

Since the Arecibo message was sent, we have pointed our transmitters at other stars and some more recently found, potentially habitable exoplanets. In 2008, the "A Message from Earth" signal was sent to Gliese 581c, only 20 light-years away, which, if you are counting on your human fingers and toes, will be received in 2028. There's a good chance we will get a reply, too, because the message itself was carefully crafted by...[checks notes]...a social media contest. Well, fuck. Never mind.

And if none of those stray signals are caught, we've also sent out physical messages in bottles. The *Pioneer plaques* were attached to space probes during the 1972 and 1973 Pioneer 10 and 11 missions. Luckily, social media didn't exist back then, so it was up to experts to devise the message, which was a dick pic and a map leading to our planet. That's right. We sent out a picture of our solar system with a big circle around Earth next to drawings of two naked people, one of which is clearly waving the universal hand gesture for "thirty-five, single male, likes long walks on the beach and anal probes."

Ghosted

Physical things travel much more slowly than radio signals. Pioneer 10 won't reach another star for over two million years. Of course, theoretically any of these messages could be intercepted in interstellar space before reaching their final destination. So, I'm sure you are dying to know, have we heard anything back yet?

In the midtwentieth century, small programs began official searches for radio-wave technosignatures. In some form or another, such efforts, which usually go by the acronym SETI (search for extraterrestrial intelligence), have been continuously monitoring the sky since then. Most recently, Yuri Milner, a multibillionaire who privately funds science projects, paid for $100 million of radio telescope time to broaden the search. Preliminary results can be described in a single word: *silence.*

Well, not complete silence—we've had our fair share of prank calls and butt dials. Take, for example, the star KIC 8462852, also called Tabby's star, named after the first author of the scientific paper that reported the discovery of its irregular light fluctuations, or sometimes also the WTF star after the subtitle of the paper, "Where's the flux?" Those sly astronomers definitely knew what they were doing. While further analysis pointed to interstellar dust clouds as the most likely explanation, that didn't stop people from suggesting the dips in light intensity were due to the harvesting of energy by alien *megastructures*. WTF, indeed.

Technosignature research is often far more speculative than biosignature research. While it often arises from legitimate questions like, "How could a technologically advanced civilization extract energy from a star with maximum efficiency?" it often quickly devolves into science fiction. Take, for example, Dyson spheres.

Dyson spheres, now the prototypical example of hypothetical megastructures, emerged from a thought experiment by physicist Freeman Dyson. Their ideal form is some encapsulation of a star entirely to harvest most, if not all, of its energy output. While Dyson thought it more as a way to illustrate the potential energy needs of advanced civilizations rather than a practical blueprint, the idea has since taken on a life of its own, presenting a tantalizing yet speculative target for technosignature searches.

The hunt for Dyson spheres and other megastructures primarily involves scanning the cosmos for signals heavy in the infrared spectrum. The idea here is that an efficient construction would extract all of the high-energy light from the star and be forced to give off lower energy as heat radiation. It's a curious blend of real science—in terms of physics calculations—and science fiction not too dissimilar from a Hollywood script. It seems we have come full circle.

Interest and funding in the search for technosignatures have fluctuated over time in perfect cadence with Earthly technological innovations. New telescopes, missions, and proposals always stir up optimism. Yet, the ghost of Fermi inevitably echoes with silence, and what was once excitement worthy of press releases becomes just another archived project folder on some institutional OneDrive.

Would robots count?

We've made the tacit assumption that aliens, whatever they might be, are *alive*—as in, they are made of living, biological stuff. Oddly, not even science fiction considers the opposite worthy of exploration. There are very few references to nonbiological aliens, and even those are obscure, like the Inhibitors from *Revelation Space*, an ancient, nonbiological intelligence seeking to prevent the advancement of civilizations across the galaxy. Their motives and origin remain shrouded in mystery, nicely highlighting the problem with imagining something so foreign. On the other hand, they are just entities with no morals, no feelings, a disdain for intelligent life, and just hell-bent on owning the galaxy's real estate. Perhaps you can name a few robots after all.

On one hand, robotic alien intelligence might be the most likely candidate. Consider the durability of machines compared to biological organisms. Robots, especially those designed for space exploration, are built to endure extreme conditions that would be lethal to organic life—radiation, vacuum, extreme temperatures, and the absence of live-streamed content. This makes them ideal explorers and settlers of the cosmos. Furthermore, the absence of a need for food, water, air, and other necessities of life, not to mention the ability to hibernate for millennia during interstellar travel, gives robots a significant advantage over organic life in the quest for spreading across the galaxy to mark our territory and cause inevitable destruction.

A particularly intriguing conceptualization of robotic E.T. is that of the self-replicating spaceship—also known as a von Neumann probe. First proposed by John von Neumann in the midtwentieth century, these machines are imagined capable of gathering materials from host planets and moons to create copies of themselves, enabling exponential growth and the potential for a civilization of machines that spreads through the galaxy much faster than biological life could ever hope to. While theoretically feasible, numerous practical hurdles prevent realistic proposals for now—which means we also haven't taken seriously the ethical challenges implied by sending our robotic slaves to ruin far-off ecosystems for no apparent purpose but unchecked expansion. Let's be honest, though. None of that is going to stop us from trying. So, why would it stop an alien civilization from doing the same?

E.T. arriving in a flying saucer suddenly feels quaint. The more likely scenario is bombardment by virulent spaceship factories crashing down on resource-rich nations without even an acknowledgment of our presence. However, the fact that exponentially self-replicating robots could "easily" occupy the galaxy is much scarier given the fact that, at least according to a little thing called evidence, they don't exist. Think about it—given the billions upon billions of years of opportunity, they should be everywhere. If they were, their destructive presence would be as noticeable as Will Smith at the 2032 Oscars. But, like all other signs of life, we haven't even seen a hint of them.

Where is everybody?

Our advances in technology have enabled us to peer deeper into the abyss, and yet, the more we see, the less we seem to find. Our once naive vision of a galactic community brimming with interstellar civilizations has sobered over time, much like a child gradually relinquishing the notion of Santa Claus, but with fewer cookies and more existential dread. We tried, Fermi, we tried.

Maybe Fermi and Drake were wrong, though. Maybe one of the terms in Drake's equation really is so close to zero that Earth is a cosmic fluke of epic proportion. This Rare Earth hypothesis posits that conditions on Earth are so uniquely suited for life that it has only happened once. In other words, "Earth-like" is not good enough; it must be "Earth-exact," and such planets are likely to be exceptionally rare. A similar concept is "the Great Filter"—note that these all must have cool names to be the subject of TED Talks and YouTube explainer videos. The Great Filter theory proposes that there's a stage in the evolutionary development of life that is incredibly hard to surpass—like a filter. Humanity might be approaching this barrier, heralding our eventual extinction, or we might be one of the very few that have passed it.

Consistent with the idea that the filter is ahead of us, perhaps civilizations are just prone to self-destruction before they are capable of interstellar travel. Maybe intelligence isn't such a good thing for the longevity of a living planet. Also, stars don't live forever. If life on Earth doesn't figure out interstellar travel, it will cease to exist in this part of the galaxy when the Sun dies in about five billion years.

This brings us to the concept of technological levels, a grading system for civilizations coined by astrophysicist Nikolai Kardashev, probably after playing too many video games. He dreamed up a scale where a Type I civilization uses all available resources on its home planet, a Type II harnesses the energy of its star (hello, Dyson spheres), and a Type III civilization of its host galaxy. By this grading, as impressed as we are with ourselves, we're not even at Type I. Maybe the boundary lies between Type I and Type II civilizations. But, if there is anything that characterizes humanity, it's blind optimism. So, as long as we are welcoming pure speculation at this point, let's get weird.

Maybe advanced civilizations are aware of us but choose to remain hidden, perhaps observing us like wildlife or avoiding contact until we reach a certain level of maturity. Do you think ants know we are watching them? It could be that Type II civilizations have advanced communication capabilities in addition to their spacefaring prowess. It's certainly not inconsistent with the laws of known physics. They might be hiding their messages with advanced techniques of encryption. I mean, radio signals? This assumes other civilizations would bother with such quaint technologies. Some have gone as far as suggesting since black holes are the most efficient storers of information, advanced civilizations will be using them as quantum computers. In fact, so the proposal goes, we should be searching for signals from black hole factories, pretending we know what those would look like.

Humanity itself uses encryption when communicating over the airwaves. To anyone but the intended target, this would look like random noise. It's been suggested that the characteristics of the signal should give it away. For example, future quantum communication will use powerful coherent laser light, which is anything but natural. On the other hand, it's been shown that quantum codes for communication and galaxy-wide computing can be hidden in starlight. So, if you still want to believe aliens are on our doorstep, just include the word *quantum* in your incoherent rambling, and no one can prove you wrong.

Existential dread by the liter

Or, maybe the answer is staring us right in the face. Even if there are other civilizations out there, the sheer scale of the universe is staggering. If some aliens on the other side of the galaxy had the ability to see Earth today, they would not find intelligent life because they would be looking at Earth as it existed 100,000 years ago.

Your sense of the scale of the universe is probably way out of whack because every picture you've seen of the solar system is wrong. Usually, the Earth and the Moon are in correct proportions—the Moon is about a quarter the Earth's size—but the Moon is placed way too close. Imagine we built a scale model of our solar system where the Earth was the size of a soccer ball. Typically, the Moon, now about the size of a tennis ball, is placed within a meter of the scale-model Earth. In reality, though, it should be placed about seven meters away, about the width of a soccer goal.

Meanwhile, the Sun would have to be a gargantuan ball the height of an eight-story building about two and a half kilometers away. If it were a clear day, you would see most of it, but the ground where it sat would be beyond the horizon. Sadly, our dear neighbor Mars would be well beyond the horizon, and the next nearest star, Proxima Centauri, would be placed in Paris, assuming our soccer game is taking place in the middle of nowhere— sorry, I'm not sorry, Missourians.

Planets like Earth are unique in that they are rare clumps of matter of extremely high density. Water, for example, has a nice round density of one kilogram per liter. Supersize your Big Mac meal, dump out whatever sugar sludge they put in the cup, and scoop up some nice, pure artisan glacial water, and you'll have one kilogram of matter in your hand. Now, imagine going to some random place in our galaxy and scooping up some Milky Way. I can't even tell you what the mass of your cup would be without using scientific notation, but it would basically be empty. If the galaxy were a uniform soup of atoms, you'd have likely scooped up a few hundred hydrogen atoms and nothing else. The point is that space is enormous and mostly empty. When our imagination shrinks the universe to fit into our brains, it becomes distorted beyond usefulness as a map of reality.

By understanding the enormity of the challenges from a scientific perspective, claims about aliens frequently visiting Earth, abducting people, or influencing ancient civilizations seem as comical as a man in a gorilla suit. Just like understanding the true nature of stars diminishes the core claims of astrology, understanding the vastness, complexity, and challenges of the universe does so for the popularized claims about extraterrestrials.

Yet...I still believe. I think there is intelligent life somewhere in the cosmos. It's probably so far away in space and time that it might as well be in a different universe. So, I won't be doing astrobiology research myself. I am glad that others are, and I will trust them for one simple reason. Science at least strives to be a public good. Scientists show us all the data and procedures to reproduce it. If one group finds a signal, others will follow and try to replicate the results. As fun as a reenactment of Barney Hill's story might be, it is just not of the same caliber.

In the end, the most profound aspect of the alien question is not in the answers we might find, but in the enduring mysteries it reveals. In our search for others, we seem to have stumbled upon the deepest enigmas of the cosmos, consciousness, and the very limits of our knowledge. In this uncharted territory, we find a boundless sea of possibility as we hear a call that beckons to the explorer in each of us. We pursue the unknown not just for what it holds, but for what it makes of us in the process. Some hear the echoes of past adventurers, scientists who have left hints and clues, while others hear the siren song of cosmic speculation.

The quest for extraterrestrial life is more than a scientific endeavor, a reflection of our deepest desires to understand our place in the universe and to connect with something beyond ourselves. Our capacity for wonder and our unyielding hope has us searching across the vast and lonely stretches of space for others looking back at us, sharing in the silent symphony of the universe.

It is thus a profound irony that you and I, dear reader, are already "aliens," foreign in many ways but also connected—not through light-years of space but through these words, evidencing something we share and could never hope to find elsewhere in the cosmos: our humanity. When a hypothetical alien holds up a mirror, supposedly reflecting an image of our intangible aspirations of cosmic significance, perhaps it really only reflects us, naked, fragile, and yearning for connection. In the vast unknown, we are unlikely to find alien intelligence, but we just may rediscover each other.

4

We are the universe's lamest time travelers

We all make mistakes. For example, I agreed to write tens of thousands of words, which forced me to research mind-numbing trash on the internet, and you paid to read them. You might benefit from a return policy, but I can't. Even so, Amazon gets its book back, and you get your money back, but there is one thing no one is getting back, especially the poor soul you forced to carry packages around town while trying to propel a barely functional e-bike. Time. That's what I was getting at.

What wouldn't you give to relive the glory days of dial-up internet and pay phones? Or maybe you crave the thrill of witnessing the birth of Jesus or the death of Genghis Khan. Perhaps it's something more practical, like revisiting your late 2021 decision to buy Bitcoin or yesterday's decision to buy Taco Bell. And why backward in time? Why not forward? Next week's lottery ticket numbers would be nice to know for the sake of your bank account balance. Come to think of it, there's an unending list of things that time travel could improve. Hell, even fast-forwarding through the current Zoom meeting you are neglecting to participate in would be a win at this point.

What if I told you that not only is time travel possible, but you are, in fact, doing it now? Every sunrise is a fresh stamp on the passport of existence, every birthday a notch on the cosmic odometer, and, depending on how much prune juice you drink, well...use your imagination. But it's the most linear, predictable, and frankly, soul-crushingly boring time travel imaginable. We shuffle forward, one sluggish second after second, through the cosmic molasses we call the present. No time-bending hot tubs, no magic telephone booths, no DeLorean-fueled joy rides back in time to flirt with the idea of incestuous paradoxes. Just us, our wrinkles deepening like temporal etchings and the gnawing awareness that the universe yawns in the face of our puny desire to be unshackled from these temporal chains.

Though the machines that purport to achieve it are new, the human fascination with time travel isn't. It's been simmering in our collective imagination since the dawn of storytellers, weaving through dusty myths, philosophical ponderings, and, eventually, whirring robots and flux capacitors. It's a dizzying ride. So, buckle up, fellow chrononauts, and don't forget the crystals because this is thirty minutes you'll never get back.

Are we there yet?

What do you think of when you hear the words *time travel*? Probably a machine that magically deposits its contents unscathed from its present time to a different time, right? This is the most straightforward and commonly depicted version of time travel in popular culture. It involves moving instantaneously from one point in the otherwise incessant flow of time to another. But this implies a concept of time about as accurate as the estimates Uber gives for the arrival of your driver.

Our intuition suggests that space is eternal and equipped with an objective clock that tracks the "true" time of the universe, and every properly functioning real clock is synchronized with it. The fictional notion of time travel is akin to turning the hands on the universal clock backward or forward. If time travel worked at all, it couldn't be like this.

Recall that Einstein gifted us with space-time, the stretchy, warpy canvas of the cosmos. We don't need to imagine all four dimensions of space and time to use it in picturing time travel. Just imagine two dimensions, one of time and one of space, like the vertical and horizontal directions on a sheet of paper. Each point on the page represents a single place and time. If we stand still, our place doesn't move, but we continue to move through time, our path tracing out a perfect vertical line in space-time. Whether we move or not, our world lines flow upward on the page, maybe veering left or right but always *forward* in time. For every observer, time seems to march along at one second per second, as it were.

When we imagine space-time as a sheet of paper, we are implicitly picking a preferred frame of reference—a preferred clock. While we cannot take that clock to be some universal one, we can take it to be the Earth's clock—or an actual clock that always remains stationary on Earth—for example. Though we can't manipulate Earth's clock, the Earth provides a pretty good frame of reference for those of us who will remain stuck to it. On the piece of paper, the naive notion of time travel is to pick up the pencil from one place in space-time and start again from another at a different point in the time dimension.

But Einstein showed us that time is *relative*, passing at different rates for clocks in relative motion. This provides us with quite a different notion of time travel. If time runs slower for me, then at some point in your future, you will find me in my past, and my present will be your future. I'll have traveled forward in *your* time. If "your" time is the same time as everyone else's on Earth, I will have done effectively the same thing as in the naive time-travel scenario. We've both moved forward in time, but a different amount has passed for each of us. While this seems very nuanced, I am bringing it up now because, surprisingly, history's first recorded notions of time travel, long before Einstein, were precisely of this relativistic variety!

Back to the future

Possibly the earliest description of time travel is found in ancient Hindu scripture, where Kakudmi inadvertently learns about the dangers of relativity because he can't find a suitable husband for his daughter Revati. Naturally, he asks his god—Brahma, who, if you recall, hatched from a golden egg to create the universe—for advice. Unfortunately, Kakudmi and Revati showed up in heaven while Brahma was enjoying some live music, so they had to wait because the track hadn't dropped on Spotify yet. What Kakudmi didn't know was that time flows more slowly in Brahma's house relative to the rest of the world. The song was short, but millions of years had passed on Earth.

Upon return, Kakudmi and Revati effectively traveled into the future. Obviously, all of Kakudmi's candidates to marry his daughter were dead, but luck would have it that another god had since incarnated himself and was totally eligible. Revati marries him, a

happy ending somewhat obfuscated by the fact that humans millions of years into the future are apparently "dwindled in stature,
reduced in vigor, and enfeebled in intellect," which necessitated
that Revati be smacked on the head with a shovel to bring her
down in size and intellect to that of an average human. It's no
Aesop fable, but I'm sure there's a moral in there somewhere.

A tale of a similar age is that of Urashima Taro, the main character of an ancient Japanese fairy tale that has been retold and
translated many times. Taro rescues a turtle that rewards him with
a trip to the Dragon Palace to be entertained by a princess. He
stays for several days but eventually gets bored because the palace
has no Wi-Fi. However, when he returns home, he discovers that
he has been gone for a hundred years. To make matters worse, he
opens a forbidden box given to him by the princess, which turns
him into an old man. The moral here is more obvious: Don't save
turtles.

These two tales highlight the relativity of time, anticipating
Einstein, at least in spirit, showing that effective time travel occurs
when one clock runs slower than another. Many other tales have
similar consequences where the slowing clock is mediated by
sleep. There's the ancient Greek story of Epimenides, who was
said to have fallen asleep in a cave for fifty-seven years, or the
Jewish tale of Honi the Circle-Drawer (long story), who fell asleep
in a cave for seventy years, or the Seven Sleepers of Christian folklore who fell asleep in—you guessed it—a cave for one hundred
ninety-eight years.

The sleep-story time-travel trope has never fallen out of favor
either. In the 1819 short story "Rip Van Winkle," Washington
Irving narrates the tale of Dutch-American Rip, who, after walking

out on his nagging wife, finds some people in the mountains getting drunk on a mysterious liquor. Naturally, he joins in, passes out, and sleeps for twenty years, waking to a world transformed by the American Revolution. Clearly embodying the American dream, Rip has become a true American hero, with statues and even bourbon named in his honor.

Using the same plot device, H. G. Wells wrote a prophetic dystopian novel, published in 1899, called *The Sleeper Awakes*. In it, Graham falls asleep and awakens two hundred years later in 2100. Social unrest, towering cities, advanced technology, blah blah, all that futuristic stuff, but with a twist. With Graham asleep and unable to spend money, his wealth had grown exponentially. Thanks to compound interest (it's what that investment influencer on TikTok was talking about), he had become the wealthiest and most powerful person in the world—sort of. I won't spoil it, but as the "good guy," Graham has both internal and external struggles, all leading to him saving the day by realizing that capitalism is bad. Who knew?

Moving on to moving pictures, sleep was obviously too boring to depict so Hollywood opted for the thaw-a-frozen-man trope, which saw its golden age in the nineties. The film *Forever Young* tells the story of Mel Gibson as a military test pilot cryogenically frozen for an unplanned fifty-three years. Upon awakening, he pretends to act but instead rapidly ages into a suspiciously accurate older version of himself. *Encino Man*...you know what...never mind. The Austin Powers series parodied the entire spy genre by following the adventures of a 1960s British spy and his archnemesis, who are both thawed in the 1990s to take advantage of obvious gags and innuendos implied by the temporally created cultural differences.

Jigo-what?

So it turns out that we didn't really even need Einstein to imagine that, with clocks moving at different speeds, time travel is natural enough not to question its use in fiction. However, you might have noticed that regardless of the relative speed of each clock in the examples so far, time still moves *forward* for everybody. It's much more difficult to imagine how one can achieve going *back* in time. Clearly, that would require some effort—specifically, "one point two one jigowatts" of effort.

Consider the most iconic "time machine"—Doc Brown's DeLorean from the Back to the Future series, which uses a "flux capacitor" powered by a billion watts, achieved through plutonium and/or a lightning strike. While a gigawatt—alternatively pronounced jigowatt by exactly one person—is a real unit of power equivalent to the output of a typical power plant, a flux capacitor is a made-up term. By the 1980s, this *technobabble*, the glorious blend of scientific jargon and sheer gibberish, had taken hold. If Doc were to describe the word-processing software I'm using to write this, he might say the following.

Doc (screaming): Great Scott, Marty! You've got to see this thing from the future! It's going to ruin everything! It's a Text Manipulation and Rendering Flogistor, and it's going to make typewriters obsolete. But the real problem is that it connects to a Global Data Nexus owned by an evil corporation, whose name is pronounced Joogle, that reads every word you write so that it can show you advertisements for things you've already bought! It's hell. Quick, grab your skateboard. We've got to stop this!

It wasn't always like this. The first known time machine was a simple clock that, when wound, ran backward and transported people nearby back in time. This fictional device appeared in a short story in 1881 called "The Clock That Went Backward" that does not describe the mechanical details of the clock using real or whimsical jargon. Though, like nearly every genre of science fiction, it is H.G. Wells who is credited for the popularization of time travel by mechanical means. In the very literally titled *The Time Machine*, the device is described in visual detail using only known science and technology elements—gadgets, dials, gears, brass, and quartz.

Some fictional devices, like the seemingly unassuming blue police box TARDIS in *Doctor Who*, have more mysterious origins. Describing TARDIS as a technologically advanced alien spacecraft capable of traversing time and space fuels our curiosity, avoids complex explanations, and embraces the universal appeal of magic portals. In *The Terminator*, a killer robot is sent from the future in a device described by a future human that followed it as "whatever it was called...the time-displacement equipment," again avoiding the *how* issue, replacing it with implied advanced science stuff lowly soldiers wouldn't understand.

Technoporn

In contrast to the whimsical and enigmatic devices of early science fiction, which portray science as an arcane domain accessible only to lone, often reckless geniuses, modern portrayals of time travel, such as those in *Interstellar* or *Avengers: Endgame*, attempt to convey "real" science by tapping our most complex theories of physics. These films illustrate a delicate balance between scientific accuracy and entertainment, navigating the thin line between making the fantastical seem plausible while attempting to avoid the complete misrepresentation of scientific realities.

In *Interstellar*, for instance, time travel is a central theme, and the film attempts to ground its story in the real scientific principles of relativity. The plot navigates the effects of gravity on time, a concept derived from Einstein's theories, which imply that time moves slower in stronger gravitational fields. This incorporation of real science adds a layer of authenticity to the story, inviting viewers to explore profound scientific concepts while being immersed in a visually stunning and emotionally resonant narrative. Unfortunately, by attributing the fifth dimension to love, which, of course, conquers everything, including logic, it quickly goes off the rails.

Avengers: Endgame approaches time travel from a comic book perspective but makes an effort to acknowledge and play with the audience's familiarity with time-travel tropes and paradoxes. The film introduces its own set of rules for time travel, using the concept of the "quantum realm," a notion loosely based on aspects of quantum physics, though heavily fictionalized to move the plot along quickly enough to fit in a half-hour final battle scene.

These modern portrayals reveal the evolving relationship between science fiction and real science. While earlier works were more inclined to sidestep detailed scientific explanations, contemporary narratives often embrace them, reflecting a growing public interest in an understanding of scientific concepts so long as they can be married to futuristic technology. However, this trend also carries inherent dangers. Making fanciful gadgets seem like tangible devices blurs the line between fiction and reality, potentially leading to misconceptions about what science can and cannot achieve.

Moreover, while these stories often feature brilliant scientists or engineers, it's crucial to avoid perpetuating the stereotype of the "lone genius." Science is a collaborative endeavor, and breakthroughs are usually the result of collective effort, not just the insights of a single individual. By balancing the wonder of discovery with a more nuanced portrayal of the scientific process, science fiction can inspire awe and curiosity without alienating those who may feel that science is beyond their grasp.

Movies and streaming shows packed with technobabble are, ironically, breeding grounds for future scientists and engineers. Sure, it inspires a sense of wonder and curiosity, perhaps sparking an interest in a career in science and technology, but it also results in a distorted understanding of scientific principles, what scientists actually do, and what it takes to be one. The disappointment is palpable on every overly keen undergraduate face when they realize Tony Stark must have spent a hundred thousand hours studying mathematics and statistics to build the Iron Man suit, which was clearly not part of the plot. Audiences take for granted the inevitability of fictional devices or theories, desperate for technological acceleration and presumably a cheap price tag, ignoring the immense complexities or straight-up impossibilities inherent in them.

Taking time to think about time

Amid all the science (fiction) and technological fantasies of time travel, there has always been a crowd creepily lurking in bushes, whispering nagging questions like, "What even is time?" Enter the philosophers.

Ancient philosophers worried mostly about the nature of time and not so much about how it could be manipulated and controlled. A prototypical example is the Greek philosopher Zeno of Elea, who presented several paradoxes of time, including "the arrow." Consider an arrow, which, paradoxically, cannot move. Why? Well, at any *instant* in time, it occupies a place and is stationary, as would have been revealed in a photo if iPhones existed twenty-five hundred years ago. The arrow cannot be moving into the place it currently occupies because it is already there, nor can it be moving to a new place because an instant has no duration. In other words, if time is entirely composed of instants, then motion itself is impossible—a classical philosophical paradox. If you know anything about philosophy, you won't be surprised that philosophers still argue about it today. Also, if you know anything about philosophy, you won't be surprised that philosophers didn't think to sort out the first problem before creating countless more.

Case in point, fourth-century philosopher Augustine of Hippo wrote that time wasn't even an objectively real thing—the past is merely a present memory, and the future is a present expectation. According to Augustine, time does not have an independent existence but is instead a function of the mind. That's deep shit. Unfortunately, the rest of the people privileged enough to own ink and quills in the following centuries were content with arguing about religious doctrine rather than sorting out interesting problems.

After the Scientific Revolution, philosophy and science became evermore intertwined. With about as much authority as one could imagine, Isaac Newton imposed strong constraints on philosophical thought. While the mathematical laws he laid down didn't have much to bear on, say, political philosophy, they greatly influenced how time was conceived and connected with equally vexing issues. Philosophers like David Hume, for example, challenged the notion of changing the past, arguing that causality, as understood by Newtonian science, would be irrevocably broken.

The discussions about the nature of time continued, fragmenting into factions like *eternalism* and *presentism*, which clash on topics such as whether the past and future are even "real" in the same sense as the present. Basically, time travel, fraught with simple and obvious contradictions, was not even worth thinking about—that is, until Einstein allowed it. In 1949, Kurt Gödel (sounds like "gow dill") found solutions to Einstein's equations of general relativity that allow *closed time-like curves*, which you can think of as loops in our paper space-time analogy. In some sense, it's not that hard to imagine when we think about reality in terms of space-time. We can go back and forth in space, so why not time as well?

Don't touch anything

Back in the nineties (best decade—I'd go back there first), I recall eagerly waiting every year for *The Simpsons* Halloween specials. Installment five, broadcast in 1994 and titled "Time and Punishment," featured Homer accidentally traveling back in time to prehistoric Springfield using a broken toaster. He recalls Grampa's warning against altering the past, "Whatever you do, don't touch anything!" Of course, Homer ignores him and swats a mosquito. Upon returning to the present, he finds himself in a dystopian society ruled by Ned Flanders and his surprising penchant for lobotomizing his slave population. Oops.

Homer goes back several times, attempting to undo his mistake, which obviously ends up worse each time. Each change Homer makes leads to a chain of causes and effects that accumulate in unpredictable ways. Since the audience was primarily twelve-year-olds, there was no need to get into the logical contradictions Homer created. But, if you watch carefully when Homer returns to the present, there is no toaster. So, clearly, he has made a change in the past that led to a future with no time machine. If there was no time machine, how could Homer have traveled to the past in the first place?

Most so-called "time-travel paradoxes" involve perverting our sacred notions of cause and effect. The most famous is the *grandfather paradox*. You've probably heard it, so I'll give you the short version. You can't go back in time to kill your grandfather because then you'd never have been born and not been able to travel back in time to kill your grandfather. Variations include killing your parents and even yourself. There are several other paradoxes that riff on this theme, all suggesting that, regardless of what Einstein had written on some paper about the physical possibility, time travel is *logically* impossible.

There are some outs. The first is simple. It assumes the past is not real, so you can't go there. Done. The next is popular because it makes for great plot devices in fiction. Called various things, such as the *Novikov self-consistency principle* or the *chronology protection conjecture*, it suggests that time, as if it has agency, prevents changes to the past in ways that would violate the laws of causality or logic. Imagine you tried to kill your grandfather in the past, but your gun jams, or you slip on a banana, or—forgetting how conception works—you were a few minutes too late, and then you curse the person who didn't call it the "*grandmother* paradox."

While this all seems amusingly inconsequential, the conclusions get existentially dark quite quickly. If you can go back in time but you can't do anything there, what does that say about free will? Arthur Schopenhauer was famously paraphrased as saying, "A man can do as he will, but not will as he will." Basically, we can do what we want, but we don't get to choose what we want. He was trying to argue that we don't have free will. But time travel makes this worse because we can't even do as we will! The possibility of changing events in the past raises the question whether our actions even matter.

However, if we can indeed influence the past or the future through time travel, it implies a level of control and unpredictability that ought to be noticeable. Surely, we would have seen these effects by now. In the most extreme case, if time travel were possible, time travelers themselves should be everywhere! But, we haven't seen a single one—or have we...

The government!

The "Philadelphia experiment" is now a term encompassing a wide range of conspiracies spanning the whole gambit—aliens, cloaking devices, teleportation, and, of course, government cover-ups. The only thing that remains consistent across its many variations is the claim that the U.S. Navy conducted a secret experiment in 1943 that caused the USS *Eldridge*, a real football field–sized warship, to disappear. The whole thing was popularized in 1984 with a cult sci-fi film by the same name that featured the "experiment" in a time-travel movie. Of course, the U.S. Navy denies such an experiment ever existed, but that's irrelevant to the true believer. Meanwhile, the person who started the rumor later admitted it was a hoax, and the ship's log shows it was never in the purported place it spontaneously disappeared from.

The saga illustrates that it only takes one idiot to launch a meme that can go viral, regardless of its veracity. Presumably, an equally attention-hungry moron connected the Philadelphia experiment to the so-called "Montauk Project," a series of top-secret government tests conducted at Montauk Air Force Station in New York. What sorts of experiments? Totally normal-sounding stuff like mind control and staging the fake Moon landing in addition to aliens and whatever else is the flavor of the month in conspiracy theorist circles. The two experiments, seemingly disconnected by decades, are related through a time warp that had sent people from the forties to the eighties. Though he passed away in 2011, one of the "survivors" has a website graciously maintained by some random internet service provider that will still sell you a $30 DVD about it if you post a check in the mail.

It's no surprise that time travel gets thrown into the mix at the breeding grounds of conspiracy theories, which have moved almost exclusively to online forums. The Time Travel Institute, which is definitely not an "institute" but an online forum boasting nearly ten thousand members, has a set of rules that are oddly more difficult to read than legal jargon. For example, rule three is as follows:

Within the Institute, each gesture, whether a comment, a shared post, or even a 'like,' creates echoes. As a chrononaut, be mindful of your digital footprints. The tools at your disposal serve to highlight wisdom or signal disruptions. Employ them judiciously, ensuring each act furthers our sanctuary's vision and spirit.

One such "chrononaut" was John Titor, the time-traveling soldier. In 2000, a person claiming to be a time traveler from 2036 appeared in the forum. He claimed he was on a military mission to retrieve an old computer from 1975 and was on a "personal" detour in the present, which was 2000. It all seemed rather quaint until he started making predictions of a future pandemic, another U.S. civil war, and World War III. With media attention coming from outside the otherwise insular internet forum full of extremely bored and occasionally insane people, the Titor story became a sensation. The forum still has posts with titles like "John Titor you still hear YOU NEeD TO REAAD THIS!"

One thing you'll save by not visiting these places is your brain cells. The theories that emerge from these cesspools are unpredictable. From claims of being the Messiah deposited into the present to blueprints for time machines made of spare auto parts, it's the Wild West of rational discourse. Most of it is nonsense, some of it is entertaining, and a sprinkling of it is actually dangerous—in a low-stakes kind of way. While specific instances of time-travel scams aren't well documented due to quick turnover and the vastness of the internet, there are some general trends and common patterns.

Consider the poor stranded chrononaut. An individual appears on a forum or social media platform claiming to be a time traveler from the future. They weave elaborate stories about how they arrived at our time but are now stuck due to a malfunctioning time machine or a depleted energy source. They then ask for financial assistance to acquire specific, often obscure, components they claim are necessary to repair their time machine.

Hey there, I'm Max from 3021. Bad news: my time pod fizzled out in your era. Good news: your primitive tech can fix it, but I'm broke in your time. Spare some change for a stranded time tourist? Promise I won't erase your timeline!

What about the classic Nigerian prince with a twist? A scammer will contact potential victims via email, claiming to have knowledge—maybe from a future version of themselves—of lottery numbers or football scores. They offer to share this information if you agree to split the profits.

Dearest Sir,

Hope this message finds you well. I'm reaching out with the opportunity of a lifetime. As incredulous as it may sound, I've been in contact with a future version of myself thanks to a minor slip in the space-time continuum...who happens to be a Nigerian prince in 2068!

Don't forget the crystals! Online listings or new-age shops in beach towns love selling gadgets claiming to enable time travel. These range from small, handheld devices to large machines, which definitely add to the credibility. The promise is that with the right knowledge or energy source, the buyer can use the device to travel through time. Such a scenario was perfectly depicted in the film *Napoleon Dynamite*, where the middle-aged Uncle Rico, yearning to relive his glory days as a high-school quarterback, buys a time machine online, a headset requiring crystals to operate that just gives him an electric shock.

Do you remember Uri Geller, the narcissistic magician who claimed he received his powers to bend spoons from aliens? Well, you won't be surprised to find out you can buy his books, which teach you how to use your "mind powers" to travel through time. Using his three rules—which, I kid you not, are 1. Focus, 2. Focus, and 3. Focus—you can revisit the past in your mind by focusing on affirmations and meditating. Congratulations, you just wasted $20 to be told what memories are in the least accurate way possible.

Ending on a more harmless example of time-travel shenanigans, in 2005, MIT students organized a "Time Traveler Convention" as a joke, inviting future time travelers to attend. It attracted quite a bit of media attention, including a brief mention on the *Saturday Night Live* news segment "Weekend Update." In the end, 450 people showed up, one in a DeLorean, but none claimed to be from the future. To be fair, apparently, it was BYOB. Lame.

How to actually time travel

At the outset, I mentioned that the relative passage of time creates effective time travel. But how exactly is that supposed to work, and is it real?

Relativistic time travel became famous, even outside of physics circles, shortly after Einstein's 1905 papers went mainstream *within* physics circles. Several scientists, including Einstein, address an apparent paradox—called the *twin paradox*—which showcases exact and calculable time travel. The simple version goes like this. There are two twins, one a homebody and the other a spacefaring adventurer. (It's not a true story.) The second twin gets into a rocket ship and flies off at close to the speed of light (for the sake of calculation, 99.9995 percent of it, say). After a year has passed on board, she decides to turn around and return to Earth, traveling at the same speed. She believes that her trip was only two years but, upon disembarking, discovers that over one hundred years have passed on Earth—that is, her twin is dead.

Upon first hearing this story, most think the "paradox" is the result that flying off near the speed of light slows down one's experience of time, but this is way off the mark. First, special relativity does not suggest that the adventurous twin *experiences* time differently. In fact, she personally experiences time passing exactly as her twin did. There is no "bullet time," as seen in *The Matrix*, predicted by relativity. Time on the spaceship really does move slower. Since all repeating processes, including biological ones, are essentially clocks, the traveling twin thinks, moves, and ages in step with every other slowed clock in her reference frame. While she can't stop time or travel back in time, she effectively travels into the future of everyone stuck on Earth.

Granted, such an experiment has never been attempted, and not because we couldn't find one half of a pair of twins to volunteer to be the potentially last human. We simply do not have anything near the technological capabilities to blast someone off on a two-year voyage into interstellar space, let alone at anything close to the speed of light. However, we did find a slightly less crazy person to give it the old college try. Astronaut Scott Kelly spent an entire year on the International Space Station between 2015 and 2016, which circles the Earth about sixteen times per day—that is, fast.

When he returned, he had aged a whopping five milliseconds less than his twin. Now, that doesn't sound very impressive, but let me remind you—Scott Kelly is an actual fucking time traveler.

There's not really a medical test one can do, like turning to the left and coughing, that can pinpoint your age down to the millisecond. Obviously, Kelly was carrying some more accurate clocks on board. We know how much time he experienced by looking at the clocks he had with him. In fact, the relativistic effects on time have been confirmed by numerous precision experiments with atomic clocks, which wouldn't lose a second of accuracy in the lifetime of the entire universe, on high-speed airplanes and satellites. May I remind you that, and not for the last time, as counterintuitive as it sounds, there is no objective definition of time. Okay, so relativistic time travel—also called *time dilation*—is possible. So, where's the paradox then?

The paradox arises because each twin sees the other as the one moving, so each should see the other's time as running slow. However, the situation is not perfectly symmetrical because the traveling twin undergoes acceleration and deceleration during her journey (turning the spaceship around to return to Earth), which breaks the symmetry. One does not *need* general relativity to explain away the paradox, but it is more enlightening because most objects are not moving at constant speeds in an unchanging direction. In fact, for clocks on satellites, which are indeed moving fast, quite the opposite effect occurs. Popular science blogs like to phrase it like this: *Your feet are younger than your head.*

Once we add gravity—or, more precisely, the effect of mass on space-time—into the picture, we find that time moves slower in reference frames near larger concentrations of mass and energy. In other words, time runs slower on Earth than in space. Your head is closer to space, so its "clock" runs faster than that in your feet. Further out in space, time runs slower due to the speed of a satellite, but that is counteracted by the gravitational effect of time dilation. The International Space Station is relatively close to Earth, but some satellites, like those that comprise GPS (the global positioning system, which we rely on to get our dumplings on time), are far enough away that time ticks much faster for them. Considering the fact that the location of your dumplings is calculated by comparing timing data to those satellites, you can imagine why accurate clocks are important. Without accounting for the time-traveling feats of those satellites, your dumplings might arrive a kilometer away instead of within a few meters of where they need to be, which is in your belly.

How to theoretically time travel

The fastest object humans have ever made is the *Parker* solar probe, which swooped close to the Sun, reaching speeds of over half a million kilometers per hour. Don't ask a Texan, but that is over a hundred times the speed of a bullet! However, it's also less than 0.1 percent of the speed of light. While it is true that we launch particles at 99.999999 percent of the speed of light in collider experiments like the Large Hadron Collider, the difference in scale and scope between that and doing the same with a human is almost too stupid to consider a serious proposition, given what we think is even technologically feasible in principle—though that hasn't stopped science fiction authors and bloggers from pretending.

The energy of a person—not even including the spaceship—traveling as fast as those particles is more than all of the world's fossil fuel reserves. Clearly, conventional modes of travel through time are out of the question. Some semi-realistic proposals exist for interstellar propulsion. For example, a *solar sail* uses radiation from stars analogous to an actual sail and wind. The wind is caused by pressure differences in air. Likewise, light, such as that from stars or even lasers, causes radiation pressure, which can be harnessed by thin reflective material. The principle has even been demonstrated by a spacecraft unironically named IKAROS, which launched in 2010 and, as of 2013, had been accelerated to about half the speed of a bullet. Solar sails have been taken seriously as modes of interstellar travel, but not at the relativistic speeds required for any meaningful sense of time travel.

Solar sails will likely not carry us, either, but perhaps our robot progeny. Recall the concept of von Neumann probes from the previous chapter. These self-replicating spaceships would require some interstellar propulsion system that, provided they were small enough, could accelerate them to a few percentage points less than the speed of light. While we grapple with the limitations of our mortal coil, these probes might silently carry on the legacy of our curiosity and ambition. In a cosmic twist of fate, as the "twins" of the original design tirelessly explore the galaxy, they might become the true inheritors of the Milky Way. To what end? Who knows. But as we are bound by the ticking of our own biological clocks, one of those inorganic twins might return to our solar system to find humanity on another planet. "It's been so long," we'll say. It obviously won't reply, but its clock will show it only just left.

Warp speed

Imagine I launched a Bluetooth speaker playing "Stairway to

Heaven" at you, traveling twice the speed of sound. Now, you'd see it coming, of course, but you wouldn't *hear* it. As soon as the speaker hit you, though, you'd start to hear the music that had been traveling behind it. However, you wouldn't hear the soothing sounds of Robert Plant, but instead some Satanic verses as the song is played in reverse because the "closest" part of the music to you is the ending.

Let's step it up a notch. Imagine I launched a Bluetooth speaker playing "Stairway to Heaven" at you, traveling twice the speed of *light*. Now, not only would you not hear it coming, but you would also not *see* it coming. When it hit you, you would actually see it moving *away* from you, all the way back to my hand. Has the speaker traveled back in time? No, not really. You hear the music in reverse, and you "see" the music in reverse, but that is just an artifact of our perception. No matter how fast I throw it, the speaker still hits you in the face *after* it leaves my hand. But, as we know, space-time has more than one trick up its sleeve.

Plot twist! The speaker was yours all along. Here's how it really went down. We were standing together, you with your speaker and me with a plan. I snatched the speaker and teleported myself a few meters away from you, then started running away very fast, but not faster than light. Shortly after, I turned around and launched the speaker back at you faster than the speed of light. What do you see? Immediately, nothing. In fact, you *still have the speaker*. In your reference frame, the speaker traveled backward in time and arrived the moment I stole it. What you see next is a sort of mirage of the speaker flying away from you and back to me, who suddenly appears to you a few meters away.

Okay, that was cute and weird, but was it problematic? Not

that version, no. However, because I'm moving away from you, events we consider simultaneous are not the same. My "now" appears to be in your past. So, if I can send an object to your location faster than light, it can arrive in your past. But I was with you in your past. If the object I sent was a bullet that killed me, how could I have shot it in the first place? We're back at the grandfather paradox. It might seem like I just made that up, but no—*if* relativity is correct, time travel is the consequence of things moving faster than light.

To cut a long story short, the vast majority of physicists do not think faster-than-light travel, or even communication, is possible. In fact, the first to hold that belief was probably Einstein himself. In his equations, he noticed that the energy of something with mass that approaches the speed of light is infinite. Clearly, then, lightspeed is an impenetrable barrier. He also pointed out the same as above—faster-than-light speeds imply the ability to communicate into the past. The only way to avoid contradictions is to "break" relativity. This is typically what science fiction writers do. Much like Newton, they assume an absolute frame of reference, which can be thought of as *narrative* time, and have the characters hop around in space, but not time.

The idea of faster-than-light travel was made famous by Star Trek's entirely fictional "warp drive." Much like many other fictional intergalactic propulsion systems, with names like hyperdrive, jump drive, or just a simple teleporter, the goal is usually to instantly travel to different places in space, with the same clock time as the original location. So, it seems we can't have relativity, faster-than-light speed, and time travel all working consistently. One has to give.

Bad science

In 1994, Miguel Alcubierre proposed a means of faster-than-light travel that appeared consistent with the equations of general relativity. Claiming to be inspired by Star Trek, he imagined contracting space ahead of the ship and expanding space behind it. Then, the ship would move forward naturally, like a cosmic surfer riding a space-time wave. Alcubierre also thought it important to note that within the little region around the ship, it doesn't need to travel faster than light. The ship just follows space-time curvature as dictated by the equations. Seemingly faster-than-light travel without any contradictions, right? If this excites you, boy, are you in luck because there are endless online forums and real-life conventions you can visit to talk about it with people who pretend to know physics.

The real problem hasn't gone away, though. In the small region of space-time around the hypothetical device, it might be moving less than the speed of light, but from the perspective of any outside observer, it is moving faster than light—which is kind of the point—and can thus be used for logically inconsistent time travel. The nice thing about these so-called "hard" science-fiction scenarios—where scientific accuracy is at least enthusiastically feigned—is that you can more easily pinpoint where the physical problem lies. In the case of warp drives, the device needs to modify the space-time around it in impossible ways. It's not that it would require an inconceivable amount of energy and just be a practical problem, but that it would require *negative* energy, which is not something that exists.

Here's an analogy. Imagine you have a beautifully detailed map of a fictional land, as you might find in a fantasy novel or video game. The map is complete with various terrains, land-marks, and paths. It appears to be a fully functional guide to nav-igating that world. However, no matter how accurate the map is, it doesn't have any practical use in the real world since what it depicts need not actually exist. You can try to use it to find your way to the grocery store, but you are just as likely to end up in a ditch.

This map, while a perfectly valid representation within its own fictional context, has no real-world application or utility. Similarly, just because negative energy exists as a mathematical extrapolation doesn't mean that it has a physical counterpart or any practical use out here in reality. It is certainly intriguing to ponder and explore theoretically, but the existence of negative energy beyond speculation is likely as elusive as Platform Nine and Three-Quarters is at the real King's Cross Station.

We've unfortunately given the name *exotic matter* to specu-lative ideas like negative energy particles or "tachyons," which have imaginary mass, allowing them to travel back in time. This reification of purely hypothetical things is problematic, making it difficult for people to differentiate between useful ideas, distant potentialities, and just plain nonsense.

Portals

Recall our piece-of-paper analogy. For the most part, we've been discussing *flat* space-time—the realm of *special* relativity. To get to general relativity—and time travel within it!—gently fold the paper back onto itself and stab a pencil through it. Congratulations, you created a *wormhole!*

Wormholes are popular in science fiction when the writers

don't want to use spaceships and engines to get the characters from one planet to the next. The portal itself is always a large circular device that, when "turned on," looks like a glassy ephemeral pool that the characters step through. The journey to the destination is often depicted identically to faster-than-light travel, tubes or tunnels like cosmic water slides spitting out their contents at the other end. You might be surprised that some of this is almost accurate—at least in theory.

Wormholes, which is a technical term for specific space-time solutions to Einstein's equations, are as old as the idea of general relativity itself. Einstein initially envisioned them as "bridges" in space-time that connect two regions in the space dimensions, allowing particles to spiral in and pop out on the other side. The goal was never to create a portal, as it were, but to "fix" the problems created by the singularity at the center of black hole solutions to his equations. It was later shown that wormholes, if they exist at all, appear and disappear so rapidly that not even light would be able to traverse them. Then, in the 1980s, Kip Thorne suggested that exotic matter (of the negative-energy-density variety, again) could stabilize the wormhole, keeping it open.

Let's pause and, just for fun, play along like everyone else on the "I'm fascinated by human space exploration" hype train. What would these wormholes actually do? First, just like in science fiction, they really would be like portals, instantly transporting you from one place in space to another. *Transporting* is not the right word, though. The wormhole takes two distant regions of space and identifies them as the same. So, you wouldn't really "move" from one to the other; they are the *same* region. That ought to look pretty weird, right? Not really, actually.

Imagine holding a glass sphere, not unlike an Instagram influencer trying to get that perfect vacation post. As you peer through it, the distant horizon appears distorted, flipped upside down, and obviously smaller. This visual distortion is analogous to how a wormhole might alter our perception of distance and connectivity in the universe. Instead of traveling across the conventional expanse of space, a journey through a wormhole would involve stepping into the sphere as your view of the other side becomes clearer, and then, without a distinct boundary or anything, you're there. If you turned around, you'd see the spherical distortion of your starting location.

Since this is technically faster-than-light travel from some observers' perspective, you won't be surprised that it allows time travel as well. By moving the "mouths" of the wormholes relative to one another, as in the twin paradox, and jumping through them in the right order, you can use them to travel forward or backward in time. Meanwhile, like the Alcubierre warp drive, nothing needs to travel faster than light in its own reference frame. More paradoxes, then? Yep.

Harry Potter and the Prisoner of Determinism

In *Harry Potter and the Prisoner of Azkaban*, there is a scene where Harry is saved by a shadowy figure casting a spell. Then, he goes back in time, where he finds his former self in trouble. To save his former self, he casts the same spell as he remembers the shadowy figure casting. Yep, it was Harry all along. Finally, Harry and the other character, whose name no one can pronounce or spell, see themselves going back in time to the exact moment they did, presumably to repeat the acts they just carried out.

Imagine again the abstract curves tracing out the paths of objects in space-time. These self-consistent time-travel trajectories are the "closed time-like curves" mentioned earlier—closed because they loop in on themselves and time-like because they don't require faster-than-light travel. They move much faster in the time dimension than they do in the space dimensions. Unlike grandfather paradoxes, which create contradictions where effects precede or even preclude their causes, closed loops appear to avoid paradoxes. However, identifying effects with their causes might create more problems than it solves.

First, let's imagine there was a "first" Harry to travel back in time to start the loop. Ah, but then who saved *that* Harry? For that matter, when did the loop start? It would seem the looping Harry has always been there, somewhat violating at least how I was told babies are made. A clearer example is just to consider a hypothetical object, like a coin. Suppose I go back in time and give myself a coin. If I live on a closed time-like curve, my younger self must then grow old and travel back in time to give the same coin to his—and my—younger self. That seems fine, but where does the coin come from? It's a single coin caught in a time loop that has no apparent origin. This is sometimes called the *bootstrap paradox*, and we can add it to the list of problems the idea of time travel faces.

If you are looking for a more personal, existential paradox, look no further than the *predestination paradox*. It's not like you have a choice in the matter. Does Harry Potter have free will? On one hand, it would certainly seem so. The entire story revolves around him making both wise and foolish decisions while he reaps the consequences. However, once he travels back in time, all that changes. At that point, it is clear that he is simply playing out a predetermined role within the closed time-like curve. If he chose not to cast the spell, he wouldn't have survived to go back in time, rendering the entire loop impossible. Therefore, Harry has no choice but to cast the spell. In other words, free will is an illusion. People love the idea of time travel, but do they know the cost?

Inventing a new reality

Unfortunately, the attitude of most who dwell on time travel is to ignore logical paradoxes...and philosophy in general. Anything is possible for physicists who want it bad enough and can convince themselves with obfuscating equations. Negative energy? Why not? Extra dimensions? Let's add seven. An entirely new theory of physics? What else is tenure for?

Hypothetical technologies like moon bases, space elevators, cold fusion, quantum computers, and so on have their skeptics. But the skepticism is always about the *practicality*, not the basic physics. These things are not ruled out by our currently best theories. Though we don't know what the engineering challenges will be to achieve them, we know they will be solved in ways consistent with existing physical laws. Warp drives and time-travel machines are not like this. Though many are conceived by practicing physicists, all require completely fictitious concepts like imaginary mass, extra dimensions, or, as noted earlier, negative energy.

The way Einstein's theory works successfully is to imagine an arrangement of real energy and matter and derive the geometry of space-time—how it looks, curves, and warps—around it. Apparently, the more fun thing to do is to invent a geometry you want and use the equations to find out what arrangement of energy and matter is required to realize it. So, if you want a wormhole or even space-time shaped like a butthole, just invent it. There's nothing stopping you. The matter and energy distribution to realize it might be illogical, but who cares? Social media trawlers are endlessly scrolling on the hunt for the next speculation from some Ivy League dipshit.

If it's not clear yet, consider the following analogy. Suppose you have 400 meters of fence to enclose a square patch of land and you need to know the area. Well, each of the four sides must be an equal 100 meters, and you can use the trusty formula that the area of a square is equal to the length of any side squared. The square with 100-meter-long sides is 10,000 square meters. Math for the win. However, now suppose you start with the demand that you need 10,000 square meters enclosed and are required to calculate how much fence you need. You can run the math backward—using the illustrious square root—to get 100 meters for the length of each side. Ah-ha—but –100 meters also works since two negatives make a positive. Does this mean we should imagine a fictitious world of negative distances existing parallel to our own world, the two of which match up in 10,000-square-meter cocaine farms? (That's what the calculation was for, by the way.) Imagine the cartel boss gave you that word problem, and you set out to Home Depot for –400 meters of fence only to come back empty-handed except for scribbles on a page detailing your compelling new theory of a "negative-sized" world.

Just because something "works" in a calculation or theory doesn't mean it bears any resemblance to the real world. Wormholes are abstract solutions to Einstein's equations that _require_ the existence of "exotic" matter—but the solutions do not actually imply the existence of any particular reality. Every worm-hole you'll ever hear about demands a solution and *then* invents "new physics" to make it work. In other words, wormholes are science fiction. They can't work in the world as we understand it.

You might think that all the cards in the time-travel game are down at this point. But the last straw any desperate theoretical physicist grabs for is the *quantum* one.

Quantum magic

To make a very long and convoluted story short, the only thing quantum physics has to say about objective reality is that observers seem to exist (that's us), and they can use equations to make statistical predictions about the outcomes of their experiments. The rest is speculative so-called "interpretation." Much like the negative-length example above, we can't help but assign meaning to the symbols in our equations.

For example, *if* you imagine that quantum particles follow trajectories in space-time, *then* you must believe that they exist in two places at once due to a phenomenon called *superposition*. However, you never had to do that in the first place. You can take the equations at face value and use them to make predictions about what you will see in experiments, which is never the trajectory of a particle. All quantum mysteries are of this flavor—first, you assume there is a classical Newtonian reality living below the quantum equations, then you arrive at nonsensical conclusions. Don't throw them out just yet, though! If you phrase them more positively as counterintuitive possibilities, then you can trade them for fake internet points.

Like the imaginary trajectories of quantum particles, it's been suggested that some quantum states can exhibit negative energy density or negative mass. The schtick is the same. Pick apart an equation in quantum physics and identify a small piece of it with a classical concept. Upon noting that it violates the classical laws of physics, suggest that you can use it for time travel. While that sounds sarcastic, it's a depressingly accurate summary of thousands of peer-reviewed academic journal articles. (It helps when your "peers" are in on the game.)

I'm not suggesting that researchers need to avoid speculative theories, but they really need to include a huge plain-language warning that the conclusions do not apply to the real world rather than rely on an implicit disclaimer absolving them from what really ought to be their responsibility to correct every idiot commenting on YouTube videos about their theories. But, no, I'm left here explaining to "chakras34" that negative quantum energy cannot reverse the effects of climate change, is not what is degrading their auras, and can't be used to power time machines built of wormholes or warp drives. Time travel is nothing more than magic that feels plausible because it is aided by symbols, formulas, and big words.

And so, despite the allure of bending the boundaries of time, the prevailing wisdom suggests that true time travel will forever remain a figment of our imagination. The laws of physics, as we currently understand them, seem to conspire against any possibility of hopping back to bygone eras or peeking ahead to future wonders. Sure, the desire to break these bounds forces us to wade through the mire of pseudoscience, scams, and conspiracy theories, but it also speaks volumes about our human ambition and curiosity, a testament to our relentless pursuit of the unattainable.

Yet, in our flights of fancy—through stories and dreams—we find a sort of liberation. These imaginary journeys into the past and future, while scientifically untenable, serve a profound purpose. They allow us to bend the rules of time—if only in the playground of our minds—offering us a glimpse of what could be in an unconstrained universe. In our quest to understand time travel, to break free from the linear march of seconds, minutes, and hours, we've unwittingly embodied the very essence of what it means to be human—to cherish the fleeting beauty of the present moment and grapple with the weight of our choices.

The irony of seeking to reclaim time, only to lose it in the pursuit, is a reminder of this. Time travel acts as a mirror, reflecting not just our desires and dreams but also the ultimate futility of seeking to outpace our own existence.

5

This is the end

As you settle into your favorite chair—you know, the one you stole from your college roommate that has also unjustifiably survived four moves and two relationships—you might glance at the calendar and wonder, "Isn't the world supposed to end today?" Wait for it... Nope, still here. Like most people, you've probably lived through at least half a dozen end-of-the-world predictions by now. Each time, the prophet confidently asserts that this time, really, the end is nigh. Yet, here we are, still paying taxes and arguing about the best brand of toothpaste—which, assuming my dentist is reading this, I totally use twice a day.

Anyway, there you are, sitting in your favorite chair, mindlessly scrolling through your social media feed as the algorithm seizes the opportunity to serve you up bread crumbs at the edge of the rabbit hole. Between bouts of existential panic and definitely not-safe-for-work content, there it is—a post shared by that one "friend" who believes lizard people run the world. He (it's always a he) declares, "The end is here! Scientists found a giant, planet-killer asteroid headed straight for Earth!...which I remind you is flat."

Well, spoiler alert: Those "scientists" aren't actually scientists but instead a guy named Gary who runs a blog from his mother's basement. Since we both know you aren't getting up, you click anyway. Gary's blog unfolds a tale of cosmic catastrophes, government cover-ups, and a secret space ark reserved for the elite. Yet, you and I are still here, totally sober, and with teeth as white as the tooth with teeth on the poster at your dentist's office. I don't know about you, but if all the government wanted from me was silence in exchange for some tax dollars and my nearly white teeth, they could have just asked nicely.

End-of-the-world conspiracies have been around since someone first looked at the sky and thought, "Fuck, that's new. Surely it means I need to sacrifice one of my children." They range from the scientifically plausible to the kind of ideas that make you wonder if sniffing glue is back in fashion. From robot overlords to zombie apocalypses, the possible ways it could all go wrong seem to be endless. But how will it *really* go down? Let me grab my trusty science stick and beat some heads and likely also some already dead horses.

The good news is that the world is not going to end soon. The bad news is also the world is not going to end soon. I mean, however it might play out, it would be spectacular to witness. Whereas I'm quite sure my own death will be uneventful. The last thing I'll experience on my deathbed is not white light at the end of a tunnel or the glorious display of a cosmic explosion but probably one of my children, sitting in the corner of my hospital room, blissfully unaware and mindlessly checking Snapchat notifications on their phone.

Never-ending stories

To understand the end, I'm afraid we must go back to the beginning—not the actual beginning, as in the big bang, but to where humans started talking about the end...so, it's more like the middle. As soon as humans evolved to form thoughts, they most certainly conceptualized at least the end of their own lives. After all, life appears to be a constant struggle to avoid one thing: death. So, once we started wasting our precious energy on idle thoughts, some individual, probably high on something foraged and staring at a campfire, wondered what happens next—and after that, and after *that*, and, well, you get the picture. Unfortunately, we can't go that far back in time to find out what the conclusion was. So, what do we actually know from recorded history about the end of history?

It's a reasonable assumption that an enormous number of end-time myths have probably existed throughout the ages of humanity, most as living, evolving oral traditions. These stories, passed down from generation to generation, would have been shaped and reshaped by the changing values, knowledge, and experiences of the people who told them. Many of these myths have been written down but, in being recorded, were likely mis-interpreted or severely lost their context. Take, for instance, the Zulu myth of the world ending with a great flood. While written records attempt to capture the essence of this myth, the original meaning and nuances conveyed in its oral form may have been altered or lost in translation, as it was written by a single person not completely immersed in the culture. Today, Zulus are pre-dominantly Christian, which clearly does not reflect their ances-tral spiritual practices.

Some myths, however, can be vetted to a degree, especially those that have been carefully preserved within their cultural con-text. An example of this is found in Aboriginal Australian culture, which is thought to have existed on the continent for tens of thousands of years. Their cultural stories have been passed down with remarkable fidelity, a fact that can be corroborated with geological evidence. End-time stories often mirror creation narratives, referring to cataclysmic events like earthquakes, tidal waves, or flooding rains.

In cases where ancient myths were discovered in written form, it's probable that these texts were merely the latest versions of much older oral stories. For example, the Norse myth of Ragnarok, eventually written down in the thirteenth century in the *Poetic Edda* and the *Prose Edda*, likely originated in an older, potentially Viking-era, oral tradition. Ragnarok is a prophesized apocalyptic event involving a great battle leading to the death of major gods like Odin and Thor, natural disasters, and the eventual submersion of the world in water.

There's a notable pattern across different cultures and times. Floods and natural catastrophes are recurring motifs, as are epic battles, such as those depicted in Ragnarok. In another example, "the longest poem ever written," the Hindu battle-myth Mahabharata unfolds a 200,000-verse tale I'm surprised someone bothered to write, let alone read. If it were written today, it would be a fifteen-second TikTok about an epic world-ending Fortnite battle that I still probably wouldn't watch.

Mark your calendars

If someone told you the world was going to end, and even went as far as detailing how it would happen, the natural question would be to ask, "So, uhh...*when?*" Unfortunately, at least for those on the receiving end of the prophecy, ancient and religious end-time myths do not have specific dates associated with them. Of course, it's clear why. Making such a specific prediction would expose the bullshi...er...prophet to a lot of risk. People would quickly stop following a leader, such as a politician, if they were constantly caught lying, right? Right? Oh well, maybe the ancients were overly cautious or not quite ready to toy with the idea of megalomania.

Now, to be fair, many ancient cultures didn't have the ability to send around end-of-the-world calendar invites. But there is a sense in which ancient prophecies came with at least vague answers to the *when* question. Most end-time scenarios play out as a sequence of events, which obviously has an opening act. For example, Ragnarok begins with the death of the god Baldr, Thor's lesser-known younger brother and the god of apparently nothing cool but looking pretty. So, although the Viking shaman couldn't predict the end of days exactly, they'd know it was imminent as soon as Baldr kicked it.

Now, if your religion has no gods left to kill, obviously your prophecy needs to have them come back to life as the ultimate second-act harbinger. All the variants in the Abrahamic tradition have such "second comings" heralding the end. The most well known—and by "well known," I mean trivialized—is from the Book of Revelation, the final book of the New Testament, which tells of the return of Jesus and some rather imaginative visions of the end times. For example, the Four Horsemen, usually taken as personifications of conquest, war, famine, and death, have been the subject of countless interpretations crossing into popular culture, including a surprising number of My Little Pony variants.

Those who attempt a literal reading of the Bible, or those who scream into loudspeakers on street corners and in mega-churches, both fear and eagerly await the Tribulation, a seven-year period of hell on earth. They fear it because the Bible tells of some pretty annoying portents in the preceding years, including serious shit like famine and disease, as well as cool shit like mass drug use leading to hallucinations of false prophets. Luckily, for true believers, Jesus will show up at the last minute and save all the Christians, who are probably a bit too enthusiastic about joining their savior in what they think will be eternal bliss. Unfortunately, most of them forget that it's not really in line with Jesus's teachings to be sadistically excited about what happens to everyone else, and will probably be left with the rest of us to enjoy all the earth-quakes, meteors, and oceans turning to blood. Only seven years? Hopefully, the drugs are still around.

In the early first century, the leader of the Jewish resistance to Roman rule declared the Messiah would arrive during the final battle, even having commemorative coins preemptively minted. The First Jewish–Roman War didn't end well, with an estimated million civilians being displaced, enslaved, or killed by the Romans. The Messiah never showed up. For the next two millennia, most popular end-time prophecies predicted specif-ically Jesus returning, which is not surprising given the wide-spread popularity of and awesome cliffhangers in Christianity. For example, Jesus said to his disciples, "Truly I tell you, this generation will certainly not pass away until all these things have happened. Heaven and earth will pass away, but my words will not pass away." I probably would have raised my hand in the back and asked, "Umm...hi. Great speech. But what do you mean by 'this generation' exactly? It's going to be important for Kool-Aid drinkers in two thousand years."

Popular interpretations of the Bible thus have many believing the Second Coming will happen in their own lifetime. As recently as 2022, 10 percent of U.S. adults believed this to be the case, while another 27 percent weren't sure. The study did not ask whether people believed Jesus was already here, though he has been spotted (allegedly) in everything from toast to Cheetos. This is all very surprising from a "fool me twice, shame on me" perspective because there is a nearly endless stream of seemingly authoritative falsified predictions that one could point to that ought to nullify the entire enterprise. From Martin Luther to Mel Gibson, religious celebrities have been wrongly predicting the Second Coming of Jesus in real life and at the box office for ages.

Predictions of this sort range from the semi-serious to batshit crazy. But whether they are written in carefully crafted prose or shouted as incoherent nonsense in the streets, you can be sure the bushes are lined with grifters and sociopathic frauds waiting to pounce. In the annals of human stupidity, we find small-scale creative swindles like the woman who wrote "Christ is coming" on her hen's eggs and shoved them back inside so she could charge visitors to watch them hatch, causing mild Judgment Day hysteria. We also discover perversions like Charles Manson's Helter Skelter and insanity like the Heaven's Gate cult, which committed mass suicide in 1997, believing they had to rapture themselves so the UFO hiding in the tail of the comet Hale-Bopp could pick up their bodies and bring them to heaven. All thirty-nine bodies were found in a recently rented house wearing identical "uniforms" consisting of a black shirt, black sweatpants, and brand-new black-and-white Nike Decades, the significance of which was, according to one surviving member, because they "just got a good deal on the shoes."

But wait...there's more

If you've taken a history class or paid attention during mass, you'll probably know that I've left out a key component to every end-time myth. If you didn't do either of those things, here's the dirty little secret about the end of days: there is no end. Despite the destruction and chaos that these prophecies bring, there's almost always a sappy sense of renewal or hope. The world, or humanity, inevitably emerges anew, often inexplicably better off than before. This cyclical view of existence is a fascinating reflection of human resilience, optimism, and gullibility. Given our penchant for continually making our environment worse, the cycle ought to feel more like an exhausting merry-go-round with unflattering lighting and distorted carnival music.

Millennia of mythology and religious doctrine have attempted to explain the purpose of human existence and the nature of reality. The explanations are invariably human-centric stories of creation, death, and rebirth. The closed cycle at least offers a sense of order to the world, but it is, at best, escapism—a convenient way to avoid confronting the reality of our own mortality.

There are some myths that reflect only the base elements of a cyclical grand narrative while remaining allegorical. For example, Hopi mythology, an oral tradition originating from what is now called Arizona, contains stories of the cyclic creation and destruction of "worlds" orchestrated by the creator Tawa. Though these stories contain a few dramatic elements, they are mostly described dispassionately. Instead of fear, the Hopi focus on acceptance of the cyclical nature of existence, striving to live in harmony with the natural world.

We briefly discussed Taoism, an ancient Chinese folk religion, with its mystical cosmic force, qi, guiding the ebb and flow of the world. The key tenet here is the pursuit of alignment within these natural cycles. In Buddhism and Hinduism, the cycle of birth, death, and rebirth has a single name: samsara. Samsara is a neutral process, not inherently good or bad, but one ultimately leading to suffering. The goal is to achieve nirvana, a transcendent state detached from material desires only temporarily achieved through anything filling in the punch line of "Sure, sex is great, but have you tried..."

While all of these cyclic cosmologies seem virtuously enlightened, the endorsed platitudes forget the realities of human nature. It's impossible for any stories to avoid moralizing interpretations or simply be used as false comfort.

Hopi is one who adheres to the "Hopi Way," which was obviously not written down as if were some contractual law but was implied by the cultural practices that allowed them to thrive in the harsh arid conditions of the high desert for however many thousands of years they had been there. Significant deviation from traditional practices is clearly apparent today, a consequence of the invasion by swathes of colonialists and retirees that look like dried raisins. Now, the subtle moralizing danger of cyclic cosmologies becomes stark. If the current world will end, and a new better one will emerge, who inherits it?

Writing referencing Taoism is often nuanced and poetic (read that as "vague and convoluted") and generously said to be untranslatable to such a vulgar language as English. If you, like me, are a monolinguist, then you will never ascend to a higher spiritual plane, which some parables suggest grants a form of immortality, conveniently misinterpreted as a get-out-of-death-free card you can purchase from the qi-synergistic guru with suspiciously bronzed skin staining his polyester robes while he leads your mediation session.

At least in the aforementioned Indian religions, such as Hinduism, there is a clear admission that the entire cycle, including rebirth, is a tedium one ought to avoid. The moral lessons aimed at avoiding world-ending calamity, as well as the entire cycle, are straightforward: achieve nirvana. It's like defeating the final boss. And, speaking of the boss, the almighty one himself only has room for those who hate all the right peopl...er...I mean, love one another...as long as it's not your neighbor's wife...unless, of course, you confess to a priest in private, but I would avoid *that* one if you catch my drift. Damn, this is hard. What does it take to get raptured over here?

Navigating the cosmic bureaucracy

Humans create stories and myths that unavoidably reflect the values of the culture in which they were created and evolved. For example, Ragnarok reflects the Norse values of strength, courage, and dragon slaying. While that seems noble enough, most end-time myths quickly morph from a poetic ending to a happily ever after post-credit scene that will only be enjoyed by those who follow a strict set of ascension rules. Anything but universal, these rules end up looking suspiciously familiar to the societies they spring from. One might be inclined to cry hypocrisy, but humans gonna human, and we apparently need rules for a sense of control, whether it is of ourselves or of others.

Sure, the Hopi tradition whispers of acceptance and harmony, but scratch the surface and you'll find echoes of their harsh desert life. Survival demands respect for water as well as a reverence for the natural cycles that bring it—think less "spiritual enlightenment" and more "annual harvest quota." Eastern mysticism, and its obsession with alignment to the cosmic order, sounds serene until you realize their ideal society was a caste system so rigid it makes the boardroom of an S&P 500 company look like a model of equitable flexibility. How convenient that the natural flow of the universe aligned perfectly with the status quo. Want to not live in squalor? That's such an Earthly desire. Sure, sharing the wealth would free you of it, but nirvana would be so much better...for both of us.

Like a dysfunctional Pokémon, Christianity has gone through numerous questionable evolutions. With all its denominations and interpretations, it resembles a divine bureaucracy, a holy hoop-jumping contest where salvation seems contingent on deciphering an ever-growing list of sacred terms and conditions. It started so simply with the Ten Commandments, a brief, inviolable to-do list for avoiding the eternal barbeque. Nowadays, Christianity seems more concerned with church attendance, tithing schedules, and gendered toilets. It's like heaven got its own homeowners' association with gates guarded by a clipboard-wielding angel named Karen.

In Greek mythology, a well-known character defied sacred traditions of hospitality, which include not killing your guests, thus failing to earn a spot in the blessed fields of Elysium. Instead, for his crimes, Sisyphus was condemned to the dungeons of Tartarus, where he was forced to roll a huge boulder up a hill only for it to roll back down before it reached the top. This picture of Sisyphus endlessly performing his futile task—now eponymously labeled *Sisyphean*—has been the subject of many interpretations over the centuries, perhaps illustrating what truly resonates with us. Basically, Sisyphus turned the otherwise moralistic mythology of the Greeks into the only sensible cyclic cosmology that simply reflects the meaningless absurdity of our pointless struggle.

This rejection of even thinking about the end was turned into an intellectual cult by cool-as-fuck philosophers like Friedrich Nietzsche and Albert Camus. Nietzsche, writing in the nineteenth century, challenged the idea of eternal life and embraced a finite existence. He advocated for living authentically, like an artist, to create our own values and meaning. Camus, a half-century later, echoed this, arguing that life has no inherent meaning. Though he didn't get the hero's welcome in the underworld, Camus suggested that Sisyphus became the true hero of humanity by accepting the absurdity of his existence. "One must imagine Sisyphus happy," Camus said.

Is that it then—should we raise a toast to the void and embrace our existential freedom by seeking out Parisian coffeehouses to hold court in with brooding looks on our faces and cigarettes hanging out of our mouths? While sitting on an intellectual throne spouting profound-sounding nonsense like "the nothing itself nothings" sounds voguish for early-twentieth-century hipsters, what modern physics has to say about the end of the world is less "lunch with Jay Gatsby at a speakeasy" and more "breakfast with Sheldon Cooper in the cafeteria"—now *that's* twenty-first-century cool.

The Big Freeze

In 1852, Lord Kelvin (not his name at the time...long story), never one to stray from scientific drama, gave us the first science-based argument for the ultimate fate of the universe. He used the new science of thermodynamics, the theory of heat, to argue that since, one, energy can't be created or destroyed, and, two, energy always dissipates through the spread of heat from hot to cold, eventually the entire universe must reach a constant temperature. In such a state, no heat can flow, no work can be done, and, essentially, nothing happens. This was later dubbed "heat death," which has confused onlookers ever since.

While catchy, the term was eventually renamed by modern-day science popularizers tired of explaining why it's a misnomer. Heat death is not *death by heat*—it's more like the death *of* heat. Heat is the flow of energy that happens when hot things come in contact with cold things. Imagine a room with a bunch of objects, including air, at different temperatures—a cup of hot coffee, a glass of cold milk, a cube of ice, and the room itself with a temperature somewhere in between. Wait long enough, and all those things will eventually come to the same temperature—the hot stuff will cool, and the cold stuff will warm. In a big enough room, the air temperature probably won't change much, but everything will reach a steady, unchanging state called *equilibrium*. Now, imagine the universe as a room—only this time, the "room temperature" is an incredibly cold -270°C. When equilibrium is reached, everything will be -270°C. Hence, the new name for the heat death of the universe is...drumroll...wait, never mind. It's already the section heading.

For Kelvin, the interesting point was not the temperature itself but that all processes in the world seem to require the exchange of heat. So, in equilibrium, nothing happens. Time itself ceases to exist. Kelvin also pointed out that since we are clearly not at equilibrium, the universe cannot be eternal—that is, it must have had a beginning. Of course, you already knew that, but recall that the big bang theory came a century after Kelvin. Since Kelvin's time, the Big Freeze scenario has only been refined, both by Einstein's general relativity and by quantum physics. Unfortunately, the outlook has become far grimmer.

In our room analogy, we imagine the end state as consisting of objects, but that's not really what happens. Our everyday experience of heat often goes unnoticed. After all, the thing we are in contact with most—the air—is completely invisible. When we think of heat, what often comes to mind is the experience of warming our hands near a fire or feeling the Sun's rays on our skin. Clearly, these things are not in contact with us. It required quantum physics to accurately explain it, but hot things also lose energy through *radiation*. The radiation is expelled as light, in most cases, not visible but infrared. We have this source of heat transfer to thank for the lovely feel of our ever-warming climate.

Even when a single atom releases light, it loses energy. Sometimes, it loses a lot of energy, transmuting into a new element entirely, expelling bits and pieces, and dangerous radiation in the form of ultraviolet and X-rays. This is called *ionizing* radiation because it can strip electrons off other atoms it comes in contact with, turning them into *ions*, which, in the case of biologically relevant atoms, is a bad thing. Luckily, universal danger symbols warn us of such sources of radiation. The radiation from the cat sleeping next to you, while radiation nonetheless, is harmless if not beneficial should you be recording it for fake internet points. Aww, so cute.

When atoms lose energy, we call it *decay*. That sounds a bit macabre, but hear me out. Energy is like a mountain. Being at the top is great, but it's a fleeting existence up there. All things must fall. We call this a "physical" law, but it's really more of a mathematical one. You see, the top of the mountain is a unique and rare place to be. It takes a lot of energy and effort to get up there. On the other hand, the "bottom" is not really a single place but many places. It comes down to simple counting in the end. There are more ways for things to be in a state with no excess energy, so that's where we expect to find them. Balls will be found in valleys, batteries will be found discharged, and atoms will all eventually decay.

Remember when your four-year-old phone's battery used to last all day? Now, it barely survives a single YouTube binge. In the morning, your battery is high on top of Mount Four Bars and One Hundred Percent Juice. Regardless of whether or not you attach it to your palm like a dopamine drug addict, it will eventually fall down the mountain to the Valley of Battery Saver Mode, a dastardly place of lies and false hope. Like an actual aging mountaineer, every time your phone scrambles over the peaks, it degrades a little, never reaching the heights it once did as promised in the advertisements conveniently served to you on last year's model as it coincidentally became sluggish. Just like the universe, your phone battery craves the ultimate low-energy state—eternal hibernation. In physics, this inevitable march to decay is called the second law of thermodynamics.

In the Big Freeze scenario, it's not that we'll end up with a bunch of "stuff" all frozen in place but that we'll all end up as a uniform sea of thinly spread particles. Unable to defy mathematics itself, every planet, every star, every intergalactic grain of dust will succumb to decay. With no usable, extractable energy left, nothing will happen, and time itself will cease to have meaning. In this lifeless void the universe will finally reveal the true extent of its pointlessness to Sisyphus—there is no mountain, no boulder, and thus not even an absurd task to do. And if that's not depressing enough, it's only what quantum physics has to say—we haven't even gotten to relativity yet.

The Big Solitude

We met space-time earlier, the sort of cosmic backdrop upon which the action plays out—with the small hitch that the action also changes the backdrop. This "warping" of space-time produces the illusion of the force of gravity as matter and energy involuntarily follow and shape the curves of space-time like a Slip 'N Slide at a frat party. Yet, as we zoom out to the big picture—the bird's-eye view, as it were—the universe is as flat as a Slip 'N Slide at a law firm's corporate retreat. Okay, technically, we only know the universe is flat within a 0.4 percent margin of error, which is exactly what Ted, a senior risk management analyst at the law firm's corporate retreat, would tell you as you all stare at the unused Slip 'N Slide watering the artificial turf in the newest partner's backyard. Ted cautions against using it.

The big bang and cosmic inflation set the universe in motion. Like a balloon being filled with air, space expands. The elasticity of the balloon counteracts this, acting like gravity pulling everything back together. There's a critical balance point where the contracting forces are just so that the balloon's expansion slows down but takes infinitely long to do so. This is what would happen to a vanilla universe in a flat space-time.

Recall the cosmological constant discussed earlier in this book. Einstein wanted his theory to predict a static universe, which was assumed to be the case at the time. Eventually, he accepted the fact that space-time was expanding, which could be accounted for with no cosmological constant (or one that is identically zero). He expressed a mild form of embarrassment over having introduced it in the first place. A famous inside joke goes that had Einstein removed his constant earlier, he would have predicted expansion and become famous.

Fast-forward to the late 1990s. Two multinational teams of astronomers studying the light from distant supernovae discovered that the expansion of the universe is not slowing down but speeding up! This posed a problem for the assumed space-time models at the time. Luckily, there was an easy fix, which was to add a constant to Einstein's equations. Oh, the irony. Had Einstein kept his cosmological constant, he could have predicted the *accelerated* expansion of the universe...making him famous! Dude just couldn't catch a break.

What is driving this additional expansion, anyway? Most scientists aren't satisfied with an answer that begins and ends by pointing at a symbol in an equation with no physical meaning. The other terms in Einstein's equations have clear meanings— curvature of space-time, energy, and mass. We don't know what the cosmological constant is, which is why it's given the mysterious name *dark energy*. You can find an endless stream of proposals from the gambit of theoretical physicists—from Nobel

Prize winners to retired engineers—for the origin of dark energy. Unfortunately, most are purely speculative nonsense, while the rest lack the compelling character to make novel testable predictions. But whatever it is, it's going to make the endgame universe much bleaker than Kelvin ever could have imagined.

There are parts of the universe that we'll *never* see, not even in principle, which sounds just plain wrong, or maybe it doesn't. It can get confusing. The Sun is eight light-minutes away. We see the light from the Sun that left eight minutes ago, from eight light-minutes away. You might expect, then, that since the universe is 13.8 billion years old, the oldest light we can see left its source that long ago. This is correct. We named this the cosmic microwave background, which was mentioned briefly in the last chapter. It was the first light in the universe. Surely, that light has traveled 13.8 billion light-years. That is also correct. The first light that left its source 16 billion light-years away from us, say, would still be traveling on its way to reach us. But it has traveled most of the journey and is only a few billion light-years away now, so we should see it soon(ish). To be pedantic, we'll see 16 billion-year-old light when the universe is 16 billion years old.

Everything in the previous paragraph is true, *assuming* the distance between every source of light and the observer is fixed. The continuous expansion of the universe complicates things. Space has expanded significantly since the big bang. In fact, the light from the cosmic microwave background is not 13.8 billion light-years away *today*. It's actually about 46 billion light-years away! Unlike objects in a convex mirror, which are closer than they appear, objects in an expanding universe are much farther away than they appear. Moreover, since everything is moving away from everything else, the farther away two things are, the faster they will be moving away from each other. At some point, the two things in question will be moving away from each other faster than the speed of light, meaning they can't even see each other. From our vantage point, there's a sphere out there, about 16 billion light-years from us, in fact, where this is happening. Any light leaving its source today beyond that sphere will never reach us.

Our observable universe, the visibility bubble centered on us, grows *and* shrinks. As time passes, we can see light that has traveled longer distances. But, as space expands, distant sources of light eventually become too far away for light to traverse the growing space between us. In the far, far future, not only will everything have decayed, but space will have expanded to the point that each particle is beyond every other's visible horizon.

A lonely electron, adrift in the endless expanse of a cold, dark universe, might begin reminiscing with a "back in my day" story, fondly recalling a nostalgic era when oppositely attracted protons were the talk of the atomic town. It would wistfully remember the thrill of the quantum dance, the buzz of photon kicks and orbital hops, only to realize, with a melancholic spin, that those days of electromagnetic attraction are just echoes of a universe that once was, now reduced to distant, unreachable memories in an ever-expanding cosmic loneliness.

The Big Rip

Since the observation of the accelerated expansion of the universe and refinements on prior astronomical observations, we settled on a "standard model" of physical cosmology widely agreed to give the simplest explanation of the data we have collected so far from the cosmos. This model contains regular matter and dark matter, which interact in various ways via the usual four fundamental forces, and dark energy, which does not. Dark energy seems to be an intrinsic property of space-time, giving it the propensity to expand. The model includes a single story for the origin of the universe in the big bang, but there are many open questions left to answer, most of which we probably have not asked yet, which renders the story of the end unwritten.

As noted, the geometry of space-time appears flat *now*. However, we can't say it has always been that way, nor that it will always be that way. If the future curvature of space-time is negative, shaped like a Pringles chip, it will expand forever without the help of dark energy. With the addition of dark energy, such a space-time will expand at an ever-increasing rate. Indeed, even in flat space-time, if the density of dark energy increases, then the expansion compounds, with distances stretching exponentially fast. This scenario, where the term *dark energy* is usually replaced with *phantom energy*, presents quite a different endgame than those previously discussed. With a name like phantom energy and just the general trend of our story so far, you won't be surprised to find out that the outlook is much, much worse.

With phantom energy, the future begins by looking not much different from the usual expansion scenario. As before, the current accelerating expansion will overwhelm the grip of gravity on the largest structures of matter as clusters of galaxies drift apart. Then, galaxies within clusters will begin to dissipate. As the exponential growth of phantom energy starts to kick in, individual galaxies will start disintegrating, their internal gravity no longer being able to hold them together against the growth of space between their stars. From our vantage point, stars will appear flung out into the void, their light stretched to near invisibility across the vast distance.

For millions of years, the Sun will be the only star visible from Earth. In roughly the same time between now and when dinosaurs roamed the Earth, our solar system will succumb. The outer planets will be the first to drift away as the rest break free from their orbital tethers, scattering into the emptiness. From Earth, the Sun will appear to shrink to the size of a typical star and then fade away altogether. Gravity will have been defeated. It's unclear if and how long anything could survive without the energy of the Sun warming the Earth. It would be a moot point, though, because in a matter of weeks, the expansion will have accelerated beyond the ability of electromagnetic forces to hold molecules together. The Earth, including everything down to the last crab, will be ripped apart from within. This kills the crab.

Lonely individual atoms will be all that's left. But, within fractions of a second, they too quickly deteriorate in the relentless expansion. As the electrons are ripped away, followed by the fracturing of atomic nuclei into subatomic particles, matter, as we know it, ceases to exist. In the final moments, even the fabric of space-time itself fails to withstand the negative pressure, tearing apart from within. At this point, called the *Big Rip*, the universe, in the truest sense of the word, ends.

The Big Crunch

Yikes. That was one ending we'll probably want to avoid. Perhaps physics has a rosier potential finale? Let's dial up the ol' end-of-days Rolodex.

We don't live in a static universe, so only two things can happen—expansion forever or eventual contraction. Expansion seems like a hard pass, so let's check in on contraction. This is the fate-of-the-universe scenario that occurs when the curvature of space is neither negative nor zero but positive. Don't get your hopes up.

In a positively curved space-time, the universe is very much like the balloon analogy. The big bang set the universe in motion, inflating the balloon against the force of gravity caused by the mutual attraction of all the mass it contains. Without dark energy or some other form of negative pressure, the momentum behind expansion eventually loses out. Recalling again that we don't know the true nature of dark energy, there is still a chance that it disappears in our apparently flat universe and expansion stops. In any case, if expansion ever does stop, *contraction* becomes inevitable. For many decades before the accelerated expansion of the universe was discovered in the late nineties, this was the assumed fate of the cosmos—the *Big Crunch*.

Much like the Big Rip, there would be many billions of years of observational evidence signaling the impending doom. With apologies to the claustrophobic, but the "doom" part should be obvious from the term *crunch*. As our technological instruments improve, we will be able to see the changing rate of expansion in more or less real time—on academic time scales, anyway. In a Big Crunch scenario, the first eye-catching thing to happen would be the rapidly increasing mergers of galaxies.

By the way, if galaxies crashing into each other concerns you, may I remind you that we are currently on a collision course with Andromeda, a galaxy the same size as our Milky Way. This is unavoidable, given their proximity, in all of the cosmological end-time scenarios, so buckle up. But fear not, as it will be less "Stranger Things" and more "Dancing with the Stars" as our two galaxies waltz around their two central supermassive black holes that attempt to suck up all the attention, which, not having seen the show, is a good analogy for its presumably washed-up celebrity host, right?

In about 4.5 billion years, Andromeda and Milky Way will start to slingshot around each other, closing in with each pass. Their large spiral arms, made of billions of stars, will whip around, sending some stars off into the void. The massive amount of interstellar gas they contain will compress, forming new stars, brightening Earth's sky with stars even a New Yorker might see. Eventually, our two black holes will merge, sending ripples through space as they release the energy of 100 million supernova explosions. Oddly enough, in this seeming violence, our solar system—even if flung from the new galaxy—will remain unscathed. Even with over a trillion stars combined, the odds of any two colliding during the merger are essentially zero. Recalling the lesson from Chapter 2, space is just unfathomably empty. In a Big Crunch universe, though, it will rapidly get more cozy.

As the distance between objects decreases, it will appear as if all the galaxies are moving toward us. But what are we "seeing" when we observe galaxies? Not only do we see visible light from stars, but they also give off high-energy radiation in the form of X-rays and gamma rays. As radiation gets compressed, its energy increases much like the pitch of an ambulance siren increases the faster it is moving toward you. Long before matter starts to crush together, the temperature will be so high as to cause nuclear fusion on the surface of stars. The Sun will essentially incinerate itself, spewing even hotter plasma into the space that surrounds it. Assuming it hasn't already spiraled into the Sun, the Earth, including everything down to the last crab, will be crushed and obliterated under the intense heat and pressure. In the last few moments, the entire universe will reach a stage similar to the beginning, an infinitely compact point of pure energy. This kills the crab.

Bounce back

As opposed to the Big Freeze and the Big Rip, the Big Crunch seems less...well, final...minus the whole crushed-to-oblivion thing. I feel like I can see you sitting there—still in that same chair but now on the edge of your seat—asking, "And then what happens?" I know you are thinking it. It's almost obvious. If the universe "ends" exactly how it began, can't it be that it just repeats itself over and over? Well, maybe, but don't start bringing those redemption vibes into it.

Imagine the *Big Bang* was preceded by a *Big Crunch*—in other words, it was a *Big Bounce!* (And I swear this is the last actual thing in cosmology that sounds like the name was stolen from a chocolate bar company.) Many have contemplated these cosmic do-overs since the dawn of nonstatic space-time. While the standard model rules out a Big Bounce altogether, a consensus was never one to stop the purest of theoretical physicists, who thrive on the fringe. Bounces have been "discovered" in many hypothetical space-time models and theories that attempt to go beyond relativity and quantum physics, a place where egos run high and evidence runs low.

Setting aside reason, there's something undeniably poetic about Big Bounces, though. They've got an appeal reminiscent of spiritual and religious cosmologies—an eternal dance of creation and destruction, a universe forever singing its own song of birth and demise. Proponents of this theory are often drawn to its aesthetic, an amusing twist in the way it mirrors the cyclical narratives found in ancient cultures.

But don't be allured by the romantic charm of these theories and their siren song masquerading as distant echoes of bygone universes. This ain't a Disney movie...unless they are producing Guillermo del Toro films now. Even if the universe repeats itself, it still gets deleted every round. The Big Crunch is still the great eraser, effectively deleting the information from past universes with the efficiency and speed of someone formatting a USB stick full of wedding photos from two marriages ago.

There's a curious asymmetry in our current understanding of the cosmos. We've got a good grasp on how it began, but even assuming the rigid structure of our theories is correct, the uncertainties in the parameters leave us with quite different possibilities. Like predicting the weather, the "near" future of our galactic neighborhood is forecastable, but the far future is not. What is "far"? Even that we don't know. But assuming the standard model is correct, which you recall leads to the Big Freeze, large black holes will still be around in a duotrigintillion years. I didn't make that number or its name up—it's 1 followed by 99 zeros, which can be written down but is impossible to imagine as a number of years. It suffices to say no one will be around to witness it. Luckily, the sequence of events leading to the end of humanity is much easier to visualize.

The final sunset

As fun as watching space-time being ripped apart or compressed to oblivion would be, it's definitely not what the last humans will see. In all likelihood, the Earth and its hardier forms of life, which have existed for billions of years, will probably gleefully witness the comparatively early demise of its most invasive species. But even the Earth itself also won't survive long enough to enjoy a slow, peaceful decay. In less than a billion years, the Sun, Earth's promise of perpetual renewable energy, will begin to show the other edge of its sword.

As it burns through its hydrogen fuel, the Sun's brightness steadily increases by about 10 percent every billion years. This might not sound like a lot, but for the surfaces of the planets that receive this additional radiation, it can be quite significant if they possess complex weather and ecosystems. Imagine the surface of the Earth, including its oceans and atmosphere, like a human body. Even though it experiences a wide range of external pressures, it has natural regulating processes to maintain its overall internal stability—so-called *homeostasis*. However, the internal processes themselves rely on being in a stable state, and leaving that zone has dire consequences. A human body can easily withstand *external* temperatures above 40°C (104°F), but once its *internal* temperature rises above that point, system failures compound. Metabolic rates increase, burning fuel and causing more heat. This promotes sweating, depriving the body of water and throwing its chemical balance off. This, in turn, leads to muscle fatigue and failure, affecting blood flow, and...well, you get the picture. Apart from the hallucinations, it's all bad.

We know what happens to human bodies experiencing high fevers because, unfortunately for those bodies, we have witnessed it happening. We don't have such detailed observations of what will happen to the Earth as its temperature rises...except, of course, we do! In the grand experiment we call *human-induced climate change,* we are witnessing in real time what happens to a warming Earth. Like a human with a fever, systems are failing. Our polar ice caps are melting, causing sea level rises. With increased evaporation rates, the water cycle is thrown off, causing more extreme weather, including more heavy rainfall and snow, which can seem paradoxical to those who feed their single brain cell with the screams of Alex Jones. We will soon find out what happens next, but all predictions suggest a continual breakdown of systems, such as the entire chemistry of the ocean as well as global weather patterns.

Now, regardless of whether you side with nearly every scientist in the world or the one wearing the tinfoil hat, the Earth will continue to warm as the Sun burns its hydrogen reserves. Life will surely survive for a billion more years through adaption and evolution, and humans may even be able to engineer solutions for our own survival by living in caves or Jeff Bezos's air-conditioned doomsday bunker. But there is absolutely nothing we can do to stop the Sun from turning Earth from a blue-green oasis to a choking desert as the relentless heat boils off the oceans, shrouding the planet in a thick, suffocating vapor atmosphere. Most multicellular life will meet its demise, though extremophiles might cling to existence in niche environments. But even these hardy survivors will find their resilience tested as the Sun enters its goblin mode phase.

Once the Sun exhausts its hydrogen fuel, it will begin to collapse as gravity wins out in the solar game of tug-of-war. However, once all that helium—the by-product of hydrogen fusion—gets squished close enough together, it too begins to fuse. This provides a temporary second life for the Sun. The only problem is that the amount of heat generated when fusing larger elements is enough to dramatically expand the Sun's outer layers. At this point, roughly five billion years from now, the Sun will have transformed from a pleasant yellow-tinted star to an angry red giant, growing to about 100 times its original size. This fiery expansion will engulf the inner planets, including Earth and its last living crab, and also possibly Mars. In this final act of solar expansion, any lingering life on Earth, no matter how resilient, will be rendered extinct as the planet is consumed by the dying Sun. This kills the crab.

The way of the dinosaur

So, apparently, both *Water World* and *Mad Max* were on the mark, though they depict scenarios billions of years apart. This kind of science fiction is branded as "postapocalyptic." But solar science shows us that no acute catastrophic events are required. The Sun is quietly carrying out its scorched-earth policy as we speak.

Now you know how the universe will end, how the Sun will die, and how the Earth will be rid of its life-giving elements. If humans make it and stick around on Earth, our story will end in a few billion years in exactly the way T. S. Elliot envisioned a hundred years ago—"not with a bang but a whimper." But even that might be "optimistic." It is estimated that over 99 percent of species that have ever existed on Earth have gone extinct. Some have gone by way of mass-extinction events, the most famous being the one that wiped out everyone's favorite prehistoric animals. As every natural history museum patron knows, about sixty-six million years ago, a 10-kilometer-wide asteroid not-so-gently landed in what is now Mexico, releasing the energy of billions of atomic bombs, basically setting the world on fire. Anything that survived the initial blast and the ensuing earthquakes, tsunamis, and acid rain would have endured the aftereffects across many, many generations. Will it happen again? Yes. Soon? Reply hazy, try again.

Currently, we know of four "near-Earth" asteroids larger than 10 kilometers wide. The most formidable among them is Ganymed 1036, spanning an intimidating 35 kilometers. If it struck Earth, it would surely be the end of all life, with the possible exception of a few of the aforementioned extremophiles. The rock itself and everything for hundreds of kilometers would be instantly vaporized. The friction alone would generate enough heat to ignite the sky in a global firestorm. Sulfur and other volatile chemicals would rain from a sky shrouded in so much debris that it blocks almost all sunlight from reaching the surface. If you are trying to imagine it by recalling some Hollywood disaster movie, don't—unless, of course, you are thinking about the Death Star.

Luckily, the big stuff in the solar system is highly visible and appears to be on stable orbits. These orbits can be observed now, and their future can be predicted by simulations. However, dynamical systems of more than two objects are *chaotic*. Chaos is what you see in a double pendulum or what is caricatured by the butterfly effect, where a butterfly flapping its wings in Tokyo causes a hurricane in Miami. Essentially, small changes now can lead to large deviations in the future. In other words, a tiny uncertainty now leads to unpredictable futures. In the far future, asteroid orbits will have changed significantly. Ganymed 1036 is expected to have an Earth-crossing orbit in about ten million years, for example.

Comets are similar to asteroids in their potential risk. Asteroids are made of rock and mostly have orbits between Mars and Jupiter, moving in the same plane and direction as the rest of the planets. They are big and dangerous, but we have pretty good tabs on them. Comets, on the other hand, are made of dirt and ice and have long, eccentric orbits that can come from any direction. They are also big and dangerous but rarely visit the inner solar system. For example, comet C/2004 R2, which grazed the Sun in 2004, has an orbital period of two billion years. On the flip side, this means there are many more comets out there on their way to visit the Sun that we don't even know about.

Historically, extinction-causing impacts hit Earth once every 100 million years or so. There is no guarantee that they will be spread apart, nor are they forbidden from bunching up in quicker succession. All we can do is estimate the odds, which are one in a million over the next hundred years. Such a chance of death is called a *micromort*, though it is typically used in reference to a single "average" person rather than all persons. As a point of reference, being alive for a single day as a twenty-year-old also exposes you to a micromort, as does smoking a single cigarette or living with a smoker for a month. Perhaps it doesn't feel so rare after all. On a positive note, the estimated warning time is at least a thousand years. Unlike the dinosaurs, we'll see something that big coming through our awesome telescopes. I'm sure society will handle its assuredly predicted doom with dignity and grace.

A bird, a plane

While a comet hitting Earth is clearly not a good outcome, comets have more traditionally been treated as signs of other forms of doom rather than being the thing that directly causes it through impact. The most famous is Halley's comet, which passes by the Earth once every seventy-six years. Throughout the ages, one can track its apparitions and ensuing hysteria. The nearest it came to Earth was in 1910. While people clearly wondered about the chances of it hitting Earth, this time, it was the comet's tail that created panic. Analysis of the light scattered off the tail demonstrated that it contained cyanogen, an extremely toxic compound. No big deal—except that calculations showed Earth was going to pass right through it! Luckily, the media—which, for the Zoomers reading this, you can think of as YouTube thumbnails but printed on paper—was there to calm the public fears...

Pfft, I could barely type that without snorting milk I didn't even drink out of my nose. Of course, they didn't do that. Instead, they printed twentieth-century clickbait with headlines such as "Comet May Kill All Earth Life, Says Scientist." Some people prayed in the streets, while others sealed up their homes. Gas-mask sales went through the roof, as did "anti-comet" umbrellas and (sugar) pills. A man in Oklahoma tried to sacrifice a virgin, hoping to ward off the comet. The weirdest fear, that the entire Earth would burn, was confessed in a letter to the Royal Observatory in Greenwich via a theory that the comet's tail was concentrated sunlight being focused through a transparent comet acting as a giant lens. In the end, the Earth didn't pass through the tail, and it wouldn't have mattered if it had. The debris blown off a comet from the Sun has a density a quintillion times less than our atmosphere, which, even if it all didn't bounce off, would contain fewer molecules than a fart.

Sadly, in 2061, when Halley's Comet returns, people will continue to do stupid things in the name of pseudoscience. It's sad because, since the dawn of the space age, we've discovered and publicly disseminated an amount of knowledge regarding comets that is comparable to the difference between traveling via jet engines and horse-drawn carriages. In fact, the last time Halley's comet passed by, in 1986, the Giotto mission performed a close flyby, providing valuable data on its nucleus and its interaction with the solar wind. A decade later, the *Stardust* spacecraft, unlike Earth, flew through a comet's tail, collected dust samples, and returned them to Earth on a capsule it dropped off as it flew by on its way to other missions. *Deep Impact* was a craft that intentionally collided with a comet to study its composition and structure. The icing on the cake was *Rosetta*, which orbited and deployed a lander on a goddamn comet, for fuck's sake. But, no, let's ignore all that and sell comet-proof salt rocks and anti-cosmic dust lotions while we host seminars on how to communicate with the comet's spirit.

Or, like Nancy Lieder, who routinely speaks to aliens, you can deny the existence of comets altogether like she did with comet Hale-Bopp in the nineties. To her, and the massive following of... how to put this nicely...morons, scientists just made up the comet's existence up a distraction from her theory that a massive object she called Nibiru or Planet X was due to collide with Earth. To anyone with a modest knowledge of astronomy or a willingness to trust someone with such knowledge, a claim like this is comically stupid. Not only would a massive planet be easily visible to astronomers, but its gravitational effects would be so obvious that even astrologers would be able to find it.

Death from within

The only other "natural" existential risk we can plausibly put odds on is a *supervolcano*. Volcanic eruptions affect local populations every year, though nowadays there is enough warning that there are typically no direct casualties. But, every century or so, a large volcano blows its top, and shit hitting the fan becomes a surprisingly apt analogy. In 1815, Mount Tambora in Indonesia erupted, directly killing tens of thousands from the cloud of 1000°C gas that traveled hundreds of kilometers per hour from the eruption and the ensuing tsunamis. It is estimated that a further hundred thousand people died globally from famine and disease as the dust blasted into the atmosphere, covered the globe, blocked sunlight, and caused the "year without a summer." Mount Tambora had a rank of seven on the *volcanic explosivity index*, which, like the more famous Richter scale for earthquakes, is logarithmic. This means that a rank eight volcano would be ten times bigger, ejecting over a quadrillion liters of lava, more than double the total amount of river water in the world.

A supervolcano might not have the ability to completely obliterate the surface of the Earth, but it would lead to mass extinction, possibly including that of its most hubristic species. Such an event is implicated in the worst historical extinction event 250 million years ago, known to geologists as the Permian-Triassic extinction and to the rest of us as the Great Dying, which sounds lovely. Obviously, the initial event would be bad for anyone nearby what with the lava, superheated gas, landslides, tsunamis, widespread fallout of pumice and ash, and all that good stuff. But it is the environmental impacts that would be global and devastating. The released chemicals would destroy the ozone layer, exposing us to increased ionizing radiation. Reduced rainfall and global cooling would devastate food production as well as, more importantly, Wall Street portfolios. While it might not be the literal end of humanity, society would collapse, mirroring the end times in several of the myths we discussed earlier, but no one is getting raptured this time.

The total existential risk of a "natural" disaster is not insignificant. Philosopher and Earth's resident disasterologist Toby Ord gives it a 1 in 10,000 chance over the next century. Some argue that, if humanity has unlimited potential, then any existential threat has an infinite cost. It's not just eight billion people that will die, but all their potential progeny should be included in the calculation as well. In other words, regardless of how unlikely each scenario is, we should probably take it more seriously. The former director of the now-closed Future of Humanity Institute, Nick Bostrom, pointed this out by noting that for every academic work on "human extinction," there were at least ten on "dung beetles" and six on "snowboarding." The lack of attention on our very real existential predicament has improved in the following decade, in part thanks to people like Bostrom, but what have we really learned?

As it turns out, getting started on planetary defense is easy if all you want to do is feel like you did *something*. Any potential threat avoidance strategy will require significant advance warning. So, the obvious first thing to do is improve our ability to detect and track stellar objects and monitor volcanic activity. Basically, we need to get better at just watching stuff, and that is indeed where most of our efforts are focused. As for what we would do if we found a threat, plenty of proposals have been made. Unfortunately, given the extraordinary nature of such an event, extraordinary proposals must be considered. And, given the ordinary nature of humans, extraordinary proposals are the only things that reach public perception. From wrapping an asteroid in aluminum foil to injecting coolant into a volcano, many strategies are better suited for science fiction. Even the once-obvious idea of blowing an asteroid up with nukes has had its efficacy questioned, not to mention the inherent risks of such an endeavor.

True to human nature, we have given ourselves a hint of optimism. On September 26, 2022, a spacecraft intentionally slammed into an asteroid, which happens to orbit another asteroid, changing its orbital period by more than thirty minutes and demonstrating that we can alter the trajectory of celestial objects. Impressed? The world was as well...sort of. The Double Asteroid Redirection Test (DART) launched recently enough to be live streamed on TikTok, boasting over 1.1 million peak viewers, slightly less than Day 22 of the Johnny Depp v. Amber Heard trial four months earlier. If you've seen the film *Don't Look Up*, you'd be forgiven for thinking it wasn't a documentary. Though, to be fair to those who couldn't care less about awesome science and the fate of the world, if we wanted to alter the orbit around the Sun of the larger of the two asteroids, we'd need to launch about a thousand simultaneous DART missions. Hopefully, that would be more exciting than a couple of bickering celebrities with careers on the cusp of irrelevance.

You do it to yourselves

Ten thousand words in, and I still haven't told you how the

world will actually end, and the real reason why billionaires are building humorously luxurious bunkers.

As science assuages our fear of natural existential threats, it also ironically produces its own through its exploitation in technological development. Let's make a long, often repeated story short. The Industrial Revolution ignited a surge in fossil fuels—coal, oil, and natural gas—used to power factories and machines. This marked a turning point, rapidly increasing the scale of emissions and setting the stage for global impacts. With the aid of scientific discoveries, fossil fuel consumption skyrocketed throughout the twentieth century, driven by factors like population growth, economic development, transportation demands, and humanity's old staple: war.

But, by the midcentury, scientific evidence began linking rising global temperatures to greenhouse gas emissions, primarily those from fossil fuels (and cow farts). At the time of writing, circa late 2023, we have increased the global average temperature by 1°C since the Industrial Revolution, and fossil fuels remain the dominant energy source globally, with around 80 percent of our energy needs met by them. Perhaps by the time you are reading this, things will have changed—but I won't be betting a reprint on it.

As this miserable century wears on, the effects of climate change are becoming increasingly evident, with rising temperatures, extreme weather events, and sea level rise posing major threats to ecosystems and an annoyance to coastal property owners. The relentless burning of fossil fuels and the resulting greenhouse gas emissions have set us on a path of environmental destabilization. This slow burn, akin to boiling a live crab by slowly increasing the temperature without it noticing, presents a danger that could fundamentally alter the conditions that make Earth hospitable for life. (You already know what happens to the crab.)

To many, a slight increase in global temperature might seem like an irrelevant or even nonproblem. After all, the temperature fluctuates wildly from day to night already, and even the shelterless people on *Alone* seem to have bigger problems than the temperature. But those brave culture warriors have simply refused to listen to the experts, who have modeled and presented the potential downstream effects countless times. These range from most-humans-will-uncomfortably-survive to oops-we-really-fucked-up.

We've already witnessed a 10-centimeter rise in ocean levels in the last thirty years. Considering that entire countries, such as the Maldives, are but one meter above the current sea level, it becomes less surprising that hundreds of millions of people and about $10 trillion in assets are at risk of disappearing in ever-increasing incidences of floods or simply the daily tides. And this is the estimate if we do nothing to mitigate the temperature rise. It's also just one obvious *direct* consequence. Remember, complex systems have interconnections that can lead to cascading failures that are impossible to predict.

Consider the 2006 European blackout, where a routine power-line disconnection led to alarms and tripped circuits that cascaded into a blackout spanning half the continent; or the 2010 Gmail crash, where nearly half of all global users lost email access because a routine software update contained a single error that cascaded through the network; or the 2021 Facebook crash, where an attempted routine maintenance request had an error that bypassed fail-safes and cascaded through the network, eventually disconnecting all of its data centers; or the 2011 Fukushima disaster, where an offshore earthquake caused a tsunami which damaged electrical grids and triggered a cascade of failures at a nuclear power plant, leading to meltdowns, radioactive releases, and long-term contamination; or...you get the picture. Basically, everything in a complex network shares its fate, and our globalized human society is connected to a complex network it shares with all the world's ecosystems and climate.

So while climate change seems like a slowly evolving issue that's always a problem for another day, cascading failures can be catastrophic even if the root cause is mundane and seemingly routine. The combined threats will add up until the tipping point is reached, and the subsequent problems can no longer be solved by more air-conditioning and finger-pointing. Economic disruption, resource depletion, food scarcity, widespread societal collapse, and potentially, the extinction of human civilization cannot be ruled out. I already noted that the ultimate fate of Earth includes climate change of the truly apocalyptic variety, but that was supposed to be hundreds of millions of years in the future, not this century. Are you ready to place your bets? Because the futurist bookies have placed 1 in 1,000 odds[1] of human extinction due to climate change happening in the next century.

Maniacs

Suffocating our civilization slowly over a few centuries is cool and all, but our ability to destroy ourselves as acutely as a comet

---

1 Odds do not include the risk of no payouts due to cash having no value in a postapocalyptic world where children are sold for toasted cockroach meals. Please gamble responsibly.

could was never considered a possibility until 1945 when we started detonating atomic bombs. The specter of nuclear warfare has since loomed over us, a testament to our capacity for self-inflicted annihilation. It's a power that elicits echoes of the anguished cry from the final and iconic scene of *Planet of the Apes*, where the protagonist time traveler finds out the "planet" is his own: "You maniacs! You blew it up! Damn you! Goddamn you all to hell!" But could we really "blow up" the world? Well, we gave it the old college try, and it seems, no, not really.

While over two thousand nuclear devices have been intentionally detonated and several nuclear reactor meltdowns have occurred, we are still here, so what's the big deal? First, let's not discredit that each detonation caused immense destruction at the blast site, with immediate casualties and long-term health effects from radiation exposure for survivors. Hiroshima and Nagasaki are stark reminders of this tragedy. Luckily, most of the detonations were relatively small tests without strategic targets or the intention to annoy Charlton Heston.

The most powerful detonation was Russia's Tsar Bomba, fifty times more powerful than the Hiroshima bomb, which was carried out in 1961 as a political bluff during the Cold War. As it was a "test," the goal was not destruction. Though it was detonated in the atmosphere above a remote uninhabited island, the effects clearly demonstrated that if all 12,512 nuclear warheads possessed by Russia, the United States, France, China, and the United Kingdom were used in the context of war, unimaginably horrific consequences would ensue.

The aftermath would play out much like that of a meteoric impact or supervolcano eruption—blast waves, thermal radiation, and fallout followed by the globe being shrouded in soot and dust, blocking sunlight, and plunging us into a "nuclear winter." As opposed to a natural disaster, where one might hope for a unified global response, a full-scale nuclear conflict kind of precludes international cooperation. Unlike natural disasters, which are impersonal forces, a nuclear war could be initiated either deliberately through the actions of national leaders, accidentally due to system malfunctions or miscommunications, or even by the whims of unstable or irrational individuals with access to nuclear launch codes. On one hand, this human element adds an additional layer of unpredictability, which has led to another 1 in 1,000 chance of an accelerated end to humanity. On the other, though, this one is entirely within our control...right?

Terminator time

No single human has sparked as much existential dread as Arnold Schwarzenegger and his army of robots dressed as naked muscle-bound Austrian clones. In the science fiction universe of the *Terminator*, Skynet, a self-aware artificial intelligence (AI) system, initiates a nuclear holocaust by launching humanity's arsenal at itself, followed by the creation of a robot army of "terminators" that hunt down the survivors and occasionally travel back in time to speak with eighties kids in thick Austrian accents. Skynet is not said to have any particular goals or values, dispassionately following its murderous programming. It's the prototypical *uncontrolled* or *rogue* AI.

Whereas HAL 9000, from Arthur C. Clarke's *Space Odyssey* and more popularly depicted in Stanley Kubrick's *2001: A Space Odyssey*, was less hell-bent on murder and more focused on its mission. HAL was an AI tasked with controlling a Jupiter-bound spaceship. Things go wrong, and the crew attempts to disconnect HAL. HAL reciprocates by killing the crew—not vindictively so, but simply because they were in the way of fulfilling HAL's goal of completing the mission. HAL had implied values that didn't include human life. It's the prototypical *misaligned* AI.

The concept of intelligent machines again dates back to ancient myths and stories, but the formal study of AI as a scientific discipline began in the midtwentieth century with its crowned "father," Alan Turing, asking the question "Can machines think?" in a formal mathematical sense. He described the main idea of what is now called *machine learning*, suggesting that computers can be *taught* through feedback given by human experts. He also suggested that the first thing an AI could do to demonstrate its thinking prowess was to beat a human at chess. He was also a prophetic genius, suggesting one would have to wait until the end of the century to see the question resolved. In 1996, IBM's Deep Blue computer defeated the reigning world champion, Garry Kasparov.

Nowadays, your smartwatch can simultaneously beat every human player at a chess tournament. But AI is not just for playing games and fodder for dystopian science fiction—it pervades your life. Autocorrect, face recognition, personalized news feeds, spam filter, product recommendations, fraud detection, chatbots, GPS navigation, self-driving cars, thermostats, voice assistants, medical image analysis, drug discovery, robotic surgery, algorithmic trading, financial risk assessment, production schedules, predictive maintenance, weather predictions, and the list goes on, are all driven by AI systems. As a modern digital citizen, you cannot avoid the use of multiple AIs—at least behind the scenes—with every action you take.

But it's this "behind the scenes" that scares some people. We have effectively lost control of AI systems. They underpin global networks of power, communications, and financial systems. They are relied upon in health care and industries at the foundation of our economy. Turning AI off—as if there were a magic switch to do so—would render your phone a paperweight and disable every computer system you use for convenience and entertainment. Recognizing how integrated AI is in modern society, one realizes these systems cannot be turned off.

That's not so scary now, but that is because each AI system is constrained to do one thing, although each does its own thing very well. Even if a program could play the perfect game of chess, it would not have the capacity to initiate a robot uprising. However, AI systems are only becoming more general and integrated. Imagine a single AI system tasked with control over all our networked systems, including financial and defense systems. It might have some programmed goal of avoiding a stock market crash and, at some point, learns that humans are the cause of such events. To achieve its mundane goal, it decides humans must be eliminated, which it can do with its access to nuclear launch sites. Although that sounds like more science fiction, recall that cascading failures are nearly impossible to anticipate in sufficiently complex systems. The only way to avoid the risk is to completely isolate AI systems, but we may well be beyond that point already.

The other approach is to ensure that AI systems are aligned with human values and intentions, working beneficially for humanity. The problem there is obvious—humans themselves are misaligned. Our wars, greed, and petty squabbles demonstrate we don't have enough understanding of actual human preferences and decision-making, let alone what makes for ideal values. If AI could figure that out, it would surely be more intelligent than us. All of the planet's existing species exhibiting advanced intelligence, such as the eight hundred Tapanuli orangutans left in the wild, are entirely reliant on the goodwill of its most intelligent species. If that were always true, do we really want to risk not being the most intelligent entities on the planet? Some envision a future where AI expands human potential, possibly even helping us negate all the previously mentioned existential threats. However, a 2022 survey of AI researchers showed that the majority believed there is a 1 in 10 chance of humanity's survival should we lose our ability to control AI.

Finally, there are pandemics. Could a virus or bacterial plague take us all out? Certainty. Are we prepared for such a scenario? Absolutely not. But this is a book about physics, not biology. The scenarios outlined here represent just a fraction of the potential dangers we know about. While popular fiction often portrays fantastical and implausible dangers like zombies, it's likely that there exist other, less sensational but equally perilous threats that are yet to capture our collective imagination. In the end, summing up all the risks, Ord reckons we have a 1-in-6 chance of total annihilation in the next century. The most positive thing to say is that this assumes we make no attempt at prevention. Indeed, some are beginning to do something about it, which is to build bunkers for themselves and a private security force. So, hit the gym, grow a beard, get some tats and cool sunglasses, pledge allegiance to capitalism, and you just might be able to ride out the apocalypse in style.

Signing off

It doesn't matter if you follow Jesus, some poorly produced YouTube videos, the instructions on your overpriced mediation app, or actual evidence—the conclusion is the same. The end is nigh. I've outlined the odds for *all* humans ceasing to exist beyond the next century, but the odds of me making it are zero. For all intents and purposes, that's the end of the world—*my* world.

There are countless ways it could all go wrong, but only one certainty. I do not fear it. *You* cannot fear it. Fear is the ultimate surrender. It's the path to stagnation, to be an engine for the destruction of the world, accelerating decay by burning useful energy. Instead, we douse the fear in the cold, hard facts of science and logic. We laugh in the face of oblivion, not with manic hysteria, but with a dry, wry chuckle born of acceptance.

Pay no attention to the advertising and false promises. Forget atonement and other cosmic bargaining chips. There is no salvation or transcendence here. Good is to carve meaning from the chaos, etching our laughter and tears onto the dissolving canvas of reality. It's the middle finger we flick at the inevitable darkness, a defiant spark in the dying embers of existence. In this grand cosmic experiment, complexity and life are fleeting. You are lucky to have it. You are lucky to *be* it. Your time is now. Do not waste it waiting for eternal bliss that will never come.

To be sure, with the apparent goal of survival, life does not want death. But, it's not the physical stuff you are currently made of that needs to survive—that changes all the time—it is the information contained within you. The *meaning* of life—your life—is to perpetuate that information. For far into the future universe, there are two possibilities to be found: the decayed remains of fleeting complexity, long forgotten and impossible to recover, or part of you—in a manner of speaking—evidencing at least a fight against decay.

Build complexity into the world—both in defiance of the end but also because it's all we have. What you make can outlast our brief dance on the precipice. Laugh in the face of this absurdity as you tell stories under the dying stars. This is not the end. We've made it to the grand finale. We may be but dust, but dammit, we tried. So, take a quiet moment before the curtain falls and raise a glass, not to forgotten gods or hollow hopes, but to the fierce, flickering flame of being itself, burning a path to oblivion.

References

CHAPTER 1

Genesis 1:1–3 (King James Version).

Main Page—LOLCat Bible Translation Project. United States, 2008. Web Archive. https://www.loc.gov/item/lcwaN0010498/.

CHAPTER 2

Centrum Jonas International. 2024. "Jonas Method." Accessed May 6, 2024. https://www.centrumjonas.com/en.

Ellis, Blake, and Melanie Hicken. 2016. "Chapter One: Who Is Behind One of the Biggest Scams in History?" CNNMoney. February 24, 2016. https://money.cnn.com/2016/02/24/news/psychic-maria -duval-chapter-one/index.html.

The Gender Experts. 2024. "Ultrasound Gender Prediction." Accessed May 3, 2024. https://thegenderexperts.com/.

Kelly, Aliza. 2024. "The Personality Traits of a Sagittarius, Explained." *Allure*, May 1, 2024. https://www.allure.com/story/sagittarius -zodiac-sign-personality-traits.

Kennedy, Stephanie. 2008. "British Secret Service Convinced of

Hitler-Astrology Link." ABC News. Last modified March 5, 2008. https://www.abc.net.au/news/2008-03-06/british-secret-service-convinced-of-hitler/1063596#.

Marco-Gracia, Francisco J. 2019. "The Influence of the Lunar Cycle on Spontaneous Deliveries in Historical Rural Environments." *European Journal of Obstetrics & Gynecology and Reproductive Biology* 236 (May): 22–25. https://doi.org/10.1016/j.ejogrb.2019.02.020.

NASA. 2012. "NASA's Hubble Shows Milky Way Is Destined for Head-on Collision." May 31, 2012. https://www.nasa.gov/mission_pages/hubble/science/milky-way-collide.html.

Novella, Steven. 2016. "Why Skepticism?" *Skeptical Inquirer* 40, no. 5 (September/October). https://skepticalinquirer.org/2017/01/why-skepticism/.

Regan, Donald T. 1988. *For the Record: From Wall Street to Washington.* San Diego: Harcourt Brace Jovanovich.

UriGeller.com. 2024. "Uri Geller." Accessed May 6, 2024. https://www.urigeller.com/.

Visontay, Elias. 2022. "'I Still Blow My Own Mind!' Uri Geller on Spoon-Bending, Showbiz and the Museum He Built to His Own Life." *Guardian*, September 27, 2022. https://www.theguardian.com/culture/2022/sep/27/i-still-blow-my-own-mind-uri-geller-on-spoon-bending-showbiz-and-the-museum-he-built-to-his-own-life.

CHAPTER 3

Ballard, Jamie. 2019. "Most Americans Believe the Government Is Hiding Info about UFOs." YouGov. July 3, 2019. https://today.yougov.com/politics/articles/24133-UFOs-government-secret-americans-poll.

Buchholz, Katharina. 2023. "Are UFO Sightings Taking Off Again?" Statista. June 30, 2023. https://www.statista.com/chart/8452/ufo-sightings-are-at-record-heights/.

Chiarelli, Nick, and Iona Kininmonth. 2022. "Global Predictions for 2023." Ipsos. December 15, 2022. https://www.ipsos.com/en/ipsos-global-predictions-2023.

de Fontenelle, Bernard Le Bovier. 1803. *Conversations on the Plurality of Worlds.* Translated by Elizabeth Gunning. London, United Kingdom: J. Cundee.

Dvali, Gia, and Zaza N. Osmanov. 2023. "Black Holes as Tools for Quantum Computing by Advanced Extraterrestrial Civilizations." ArXiv. Last modified November 28, 2023. https://doi.org/10.48550/arXiv.2301.09575.

NASA Exoplanet Science Institute. "NASA Exoplanet Archive." Caltech. Accessed May 6, 2024. https://exoplanetarchive.ipac.caltech.edu/.

National Archives. 2020. "Project BLUE BOOK—Unidentified Flying Objects." Last updated September 29, 2020. https://www.archives.gov/research/military/air-force/ufos.

Orth, Taylor. 2022. "A Growing Share of Americans Believe Aliens Are Responsible for UFOs." YouGov. October 4, 2022. https://today.yougov.com/technology/articles/43959-more-half-americans-believe-aliens-probably-exist.

Scharf, Caleb A. 2019. "The First Alien." *Scientific American.* November 23, 2019. https://blogs.scientificamerican.com/life-unbounded/the-first-alien/.

"The Lucian of Samosata Project." The Lucian of Samosata Project. Accessed May 3, 2024. http://lucianofsamosata.info/wiki/doku.php.

Vergun, David. 2023. "DOD Working to Better Understand, Resolve Anomalous Phenomena." U.S. Department of Defense. April 29, 2023. https://www.defense.gov/News/News-Stories/Article /Article/3368109/dod-working-to-better-understand-resolve -anomalous-phenomena/.

CHAPTER 4

Bielek, Alfred. "The Philadelphia Experiment & Montauk Survivor Accounts" (DVD). Accessed May 3, 2024. http://www.bielek.com /order.htm.

Impey, Chris. 2021. "Have We Made an Object That Could Travel 1% the Speed of Light?" The Conversation. November 15, 2021. https://theconversation.com/have-we-made-an-object-that -could-travel-1-the-speed-of-light-170849.

"Philadelphia Experiment." 2015. *The Skeptic's Dictionary* (blog). Last modified November 21, 2015. https://skepdic.com/philadel .html.

Pilkington, Mark. 2005. "Do the Time Warp." *Guardian.* June 8, 2005. https://www.theguardian.com/science/2005/jun/09/farout.

Schladebeck, Jessica. 2016. "A Look at 'Stranger Things' and the Secret Government Experiments That Inspired It." *New York Daily News.* September 1, 2016. https://www.nydailynews.com/2016 /09/01/a-look-at-stranger-things-and-the-secret-government -experiments-that-inspired-it/.

Smith, Nicholas J.J. 2024. "Time Travel." Edited by Edward N. Zalta & Uri Nodelman. *Stanford Encyclopedia of Philosophy.* Last modified March 22, 2024. https://plato.stanford.edu/entries/time-travel/.

Time Travel Institute. 2012. "Rules & Guidelines." May 2, 2012. https://timetravelinstitute.com/pages/guidelines/.

UriGeller.com. "Instant Meditation." Accessed May 3, 2024. https://www.urigeller.com/instant-meditation/.

CHAPTER 5

All Out Attack. 2020. "Going to Heaven on a UFO—an Interview with Heaven's Gate Cult Members 22 Years On." January 20, 2020. https://alloutattackuk.wordpress.com/2020/01/20/5334/.

Bostrom, Nick. 2013. "Existential Risk Prevention as Global Priority." *Global Policy* 4, no. 1 (February): 15–31. https://existential-risk .com/concept.

Boyd, Matt, and Nick Wilson. 2023. "Assumptions, Uncertainty, and Catastrophic/Existential Risk: National Risk Assessments Need Improved Methods and Stakeholder Engagement." *Risk Analysis* 43, no. 12 (December): 2486–2502. https://doi.org/10.1111/risa .14123.

Carlton, Genevieve. 2021. "The Story of the Apocalyptic Frenzy Inspired by the Arrival of Halley's Comet in 1910." Edited by Erik Hawkins. *All That's Interesting.* August 23, 2021. https://allthatsinteresting.com/halleys-comet-1910.

Diamant, Jeff. 2022. "About Four-in-Ten U.S. Adults Believe Humanity Is 'Living in the End Times.'" Pew Research Center. December 8, 2022. https://www.pewresearch.org/short-reads/2022/12/08 /about-four-in-ten-u-s-adults-believe-humanity-is-living-in-the -end-times/.

Intergovernmental Panel on Climate Change. 2018. "Special Report: Global Warming of 1.5°C." Accessed May 6, 2024. https://www .ipcc.ch/sr15/.

Jägermeyr, Jonas, Alan Robock, Joshua Elliott, Christoph Müller, Lili Xia, Nikolay Khabarov, Christian Folberth, et al. 2020. "A

Regional Nuclear Conflict Would Compromise Global Food
Security." Edited by Christopher B. Field. *Proceedings of the
National Academy of Sciences* 117, no. 13 (March): 7071–7081.
https://doi.org/10.1073/pnas.1919049117.

Josephy Jr., Alvin M. 1973. "The Hopi Way." *American Heritage* 24, no.
2 (February). https://www.americanheritage.com/hopi-way.

Michel, P., R. Gonczi, P. Farinella, and Ch. Froeschlé. 1999.
"Dynamical Evolution of 1036 Ganymed, the Largest Near-
Earth Asteroid." *Astronomy and Astrophysics* 347 (July): 711–719.
https://ui.adsabs.harvard.edu/abs/1999A%26A...347..711M/
abstract.

NASA. "Understanding Sea Level: By the Numbers." NASA Sea
Level Change. Accessed May 6, 2024. https://sealevel.nasa.gov
/understanding-sea-level/by-the-numbers.

Ord, Toby. 2020. *The Precipice: Existential Risk and the Future of
Humanity.* New York: Hachette Books.

Ruth, Kent. 1984. "Sacrificial Virgin Saved Just in Time, Story Says."
The Oklahoman. September 9, 1984. https://www.oklahoman.com
/story/news/1984/09/09/sacrificial-virgin-saved-just-in-time
-story-says/62791072007/.

Salotti, Jean-Marc. 2022. "Humanity Extinction by Asteroid Impact."
Futures 138 (April): 102933. https://doi.org/10.1016/j.futures
.2022.102933.

Acknowledgments

I thank my fans, especially those of *Quantum Bullsh*t*, who have contacted me with words of praise and encouragement. This book is for you!

I thank Lindsay Ferrie, as always, for being my hype woman, tirelessly reading drafts left on her pillow, and just putting up with me in general.

I thank my children for taking care of me when I'm older. (It's in print now, suckers!)

I thank my friends and family who have bounced back ideas, not just on this book, but many things.

I thank Geraint Lewis for first suggesting the idea of a "cosmic sequel" to *Quantum Bullsh*t*. We were supposed to write this together, but your replies to my emails were too slow. Consider yourself scooped.

Last, but definitely not least, I thank my editor, Anna Michels, and her team. While so many editorial software tools exist today, nothing can replace someone who is such an expert at her craft.

About the Author

Chris Ferrie is an associate professor at the University of Technology Sydney in Australia, where he researches and lectures on quantum physics, computation, and engineering. He is the author of *Quantum Bullsh*t: How to Ruin Your Life with Quantum Physics* and *42 Reasons to Hate the Universe (And One Reason Not To)*. He also happens to be the number one bestselling science author for kids. Though those books have decidedly fewer f-bombs.

Did you love
Cosmic Bullsh*t?

Don't miss any of bestselling science writer
Chris Ferrie's hilarious, no-bullsh*t books
about the universe, quantum physics,
and more brain-bending topics!

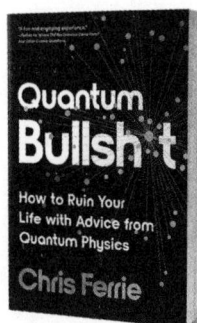

Quantum
Bullsh*t.
How to Ruin Your
Life with Advice from
Quantum Physics
Chris Ferrie

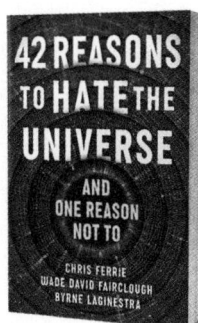

42 REASONS
TO HATE THE
UNIVERSE
AND
ONE REASON
NOT TO
CHRIS FERRIE
WADE DAVID FAIRCLOUGH
BYRNE LAGINESTRA

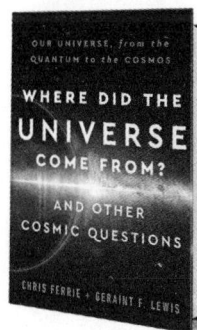

OUR UNIVERSE, from the
QUANTUM to the COSMOS
WHERE DID THE
UNIVERSE
COME FROM?
AND OTHER
COSMIC QUESTIONS
CHRIS FERRIE + GERAINT F. LEWIS

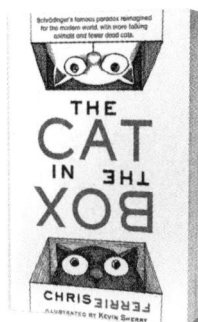

THE
CAT
IN THE
BOX
CHRIS FERRIE
ILLUSTRATED BY KEVIN SHERRY